# Object Oriented
# Software
# Technologies in
# Telecommunications

# Object Oriented Software Technologies in Telecommunications

## From theory to practice

Edited by

**Iakovos Venieris**
*National Technical University of Athens, Greece*

**Fabrizio Zizza**
*Siemens Information and Communication Networks SpA, Italy*

**Thomas Magedanz**
*IKV++ GmbH Informations und Kommunikations Technologie, Germany*

**JOHN WILEY & SONS, LTD**
Chichester · New York · Weinheim · Brisbane · Singapore · Toronto

Copyright © 2000 by   John Wiley & Sons, Ltd
                      Baffins Lane, Chichester,
                      West Sussex, PO19 1UD, England

                      National  01243 779777
                      International  (+44) 1243 779777
e-mail (for orders and customer service enquiries): cs-books@wiley.co.uk

Visit our Home Page on http://www.wiley.co.uk or http://www.wiley.com

*Other Wiley Editorial Offices*

New York • Germany • Brisbane • Singapore • Toronto

*Library of Congress Cataloging-in-Publication Data*

Object oriented software technologies in telecommunications : from theory to practice / edited by Iakovos S. Venieris, Fabrizio Zizza, Thomas Magedanz.
     p. cm.
   Includes bibliographical references and index.
   ISBN 1-471-62379-2 (alk. paper)
     1. Telecommunication—Computer programs. 2. Object-oriented programming (Computer science) 3. Software engineering.   I. Venieris, Iaskovos. II. Zizza, Fabrizio. III. Magedanz, Thomas.

   TK5102.5. O184 2000
   621.382'0285'—dc21

                                                              00-023094

*British Library Cataloguing in Publication Data*

A catalogue record for this book is available from the British Library

ISBN  0 471 62379 2

Produced from PostScript files supplied by the authors.

Printed and bound in Great Britain by CPI Antony Rowe, Eastbourne

# Contents

# Preface

The face of telecommunications is changing. Telecommunication services are becoming increasingly complex, covering advanced multi-media, multi-party services, and involving increasingly personal and terminal mobility. The underlying network is made up of ever more heterogeneous technologies, including fixed and wireless networks using both circuit switched and packet oriented transmission mechanisms. With progressing deregulation of the telecommunications markets, networks and services are being operated by an increasing number of co-operating and competing service and network providers. The key to managing these three dimensions is *programmability*. Making networks 'programmable' enables service and network providers to deploy a wide range of differentiated services in a heterogeneous network environment while keeping it simple to create, manage and use them.

One way of achieving programmability of telecommunication networks is by introducing a *middleware* that hides the technology specifics of the communication equipment from the applications that use it. As the name suggests, middleware sits in the 'middle' between the operating system and the applications. It encapsulates a number of functions that are not part of the operating system, but that are common to many applications, like sending messages from one component to another. These middleware platforms were based on state-of-the-art information technologies. In the first half of the nineties these were distributed object technologies, whereas in the second half, machine independent code and mobile agent technologies gained considerable momentum. Now, at the turn of the century, integrated middleware solutions based on both distributed objects and mobile agents are considered as the best basis for realizing flexible open service platforms. As middleware typically provides only lower level generic functionalities, another prerequisite for achieving the rapid and efficient provision of telecommunication services in the above-mentioned context are *open service architectures*, which provide frameworks for the definition, combination and use of both generic and specific low to high level service components.

The first combined middleware and open service architecture in the context of telecommunications represents the *Intelligent Network* (IN), which was defined in the middle of the eighties. The IN provides a network and service independent architecture, which is primarily based on the incorporation of state-of-the-art information technology into the telecommunications world. Whereas the traditional centralized IN architectures are based on the Remote Procedure Call (RPC) paradigm, and the availability of common channel signaling networks in the seventies, today distributed IN architectures have become possible through the application of distributed object technologies and mobile agent technologies.

Based on these considerations, this book is organized into three main parts, with each part comprising several chapters.

Part I: *The need for Advanced Software Technologies in Telecommunication Networks* describes the changing face of the telecommunications environment, driven by regulatory, market and technological issues. In particular, the need for network programmability will be motivated, leading to the need of advanced telecommunications middleware platforms based on state-of-the-art software technologies.

Part II: *Enabling Software Technologies* provides a tutorial type introduction to the emerging software technologies, comprising object-oriented design methodologies, distributed object technologies, machine independent code and intelligent and mobile agent technologies.

Part III: *Case study: Distributed Intelligent Broadband Networks* illustrates the impacts of the previously outlined software technologies on the evolution of Intelligent Networks. Starting from the given limitations of the traditional IN concept and architecture, this part outlines the main evolution path towards a distributed IN for broadband networks, and provides a detailed overview of the broadband IN architecture based fundamentally on the application of both distributed object and mobile agent technologies in an integrated manner.

This book is a common product of a large group of authors working in several European countries within the international project *MARINE (Mobile Agent EnviRonments in Intelligent NEtworks)*, which was sponsored by the European Union as part of the ACTS (Advanced Communication Technologies and Services) research program. This project was a unique opportunity for a group of scientists and developers from several industrial companies and research institutions to analyze the potentialities of emerging software technologies in the context of next generation Intelligent Broadband Network implementation in significant detail.

The result, documented in this book, is on the one hand a tutorial on the state-of-the-art software technologies of major relevance for next generation telecommunications (this is the theory). On the other hand, the book provides a consistent set of architecture definitions, service descriptions, and interface and protocol specifications based on these software technologies, which have proven their practical viability through a prototype implementation and a trial at the end of the project involving real users. We hope that readers will follow this journey from theory to practice and enjoy the reading.

## Acknowledgements

The editors would like to thank all members of the MARINE project for their excellent work. In particular, we would like to thank those colleagues who have contributed to this book significantly but who are not mentioned as authors, namely

Bernhard Bamberg, T-Nova Deutsche Telekom Innovationsgesellschaft mbH
Donatella Blaiotta, Siemens Information and Communication Networks S.p.A.
Jose Maria Cavanillas, Sema Group sae
Sergio Daneluzzi, Siemens Information and Communication Networks S.p.A.
Mark Donnelly, Teltec Ireland

Charles Francis, Swisscom
Ivan Griffin, Teltec Ireland
Lorenz Lehmann, Swisscom
Christoph Maeder, Swisscom
Conor Morris, Teltec Ireland
Wolfgang Mueller, T-Nova Deutsche Telekom Innovationsgesellschaft mbH
Elena Orlandi, Siemens Information and Communication Networks S.p.A.
Ana Pinuela, Sema Group sae
Spyros Polykalas, National Technical University of Athens
Anne-Marie Sassen, Sema Group sae
Vassilios Stathopoulos, National Technical University of Athens

Furthermore, we would like to thank the publisher for the effective and friendly guidance
they provided for this work.

Orbix and OrbixWeb are registered trademarks of IONA Technologies; VxWorks and
Tornado JVM are registered trademarks of Wind River Systems; Solaris and Java are
registered trademarks of Sun Microsystems; Microsoft Windows is a registered trademark
of Microsoft Corporation. All other trademarks are property of their registered owners.

**Iakovos Venieris**
National Technical University of Athens, Greece

**Fabrizio Zizza**
Siemens Information and Communication Networks spa, Italy

**Thomas Magedanz**
IKV++ GmbH, Germany

# Contributors

**Markus Breugst**
*IKV++ GmbH*
*Berlin*

**Fotis Chatzipapadopoulos**
*National Technical University of Athens*
*Athens*

**Sang Choy**
*IKV++ GmbH*
*Berlin*

**Giovanna De Zen**
*Siemens Inform. And Comm. Networks S.p.A*
*Milan*

**Lorenzo Faglia**
*Italtel S.p.A.*
*Milan*

**Thomas Magedanz**
*IKV++ GmbH*
*Berlin*

**George Mamais**
*National Technical University of Athens*
*Athens*

**Gennaro Marino**
*Italtel S.p.A.*
*Milan*

**Andreas Papadakis**
*National Technical University of Athens*
*Athens*

**Menelaos Perdikeas**
*National Technical University of Athens*
*Athens*

**Odysseas Pyrovolakis**
*National Technical University of Athens*
*Athens*

**Iakovos Venieris**
*National Technical University of Athens*
*Athens*

**Fabrizio Zizza**
*Siemens Inform. And Comm. Networks S.p.A*
*Milan*

# Acronyms and Abbreviations

**AAL5**   ATM Adaptation Layer 5
**AC**   Agent Communication
**ACC**   Agent Communication Channel
**ACL**   Agent Communication Language
**ADSL**   Asymmetric Digital Subscriber Line
**AEE**   Agent Execution Environment
**AMS**   Agent Management System
**ANM**   Answer Message
**AO**   Application Object
**AP**   Agent Platform
**API**   Application Programming Interface
**ARPA**   Advanced Research Projects Administration
**AS**   Access Signaling
**ASN.1**   Abstract Syntax Notation 1
**ATM**   Asynchronous Transfer Mode
**AWB**   Aglets Workbench
**BA**   Bootstrapping Agent
**B-CCF**   Broadband Call Control Function
**BCM**   Basic Call Model
**BCP**   Basic Call Processing
**BCSM**   Basic Call State Model
**B-INAP**   Broadband Intelligent Network Application Protocol
**B-IP**   Broadband Intelligent Peripheral
**B-ISDN**   Broadband Integrated Services Digital Network
**BOA**   Basic Object Adapter
**BRI**   Basic Rate Interface
**B-SCF**   Broadband Service Control Function
**B-SCP**   Broadband Service Control Point
**B-SDF**   Broadband Service Data Function
**B-SRF**   Broadband Specialized Resource Function
**B-SS&CP**   Broadband Service Switching and Control Point
**B-SSF**   Broadband Service Switching Function
**B-SSP**   Broadband Service Switching Point
**BVT**   Broadband Video Telephony
**CBR**   Constant Bit Rate
**CCAF**   Call Control Agent Function

**CCF**    Call Control Function
**CGI**    Common Gateway Interface
**CLASS**    Customer Local Area Signaling Services
**CMIP**    Common Management Information Protocol
**CMIS**    Common Management Information System
**CoD**    Code on Demand
**COM**    Component Object Model
**CORBA**    Common Object Request Broker Architecture
**CPE**    Customer Premises Equipment
**CPU**    Central Processing Unit
**CS**    Capability Set
**DAE**    Distributed Agent Environment
**DAI**    Distributed Artificial Intelligence
**DARPA**    Defense Advanced Research Projects Agency
**DCA**    Data Connection Adapter
**DCE**    Distributed Communications Environment
**DCOM**    Distributed Component Object Model
**DF**    Directory Facilitator
**DIBN**    Distributed IBN
**DII**    Dynamic Invocation Interface
**DNS**    Domain Name System
**DOT**    Distributed Object Technology
**DP**    Detection Point
**DPE**    Distributed Processing Environment
**DS0**    Digital Signal 0 - 64 Kbps channel
**DSI**    Dynamic Skeleton Interface
**DSM-CC**    Digital Storage Media Command and Control
**DTE**    Data Terminal Equipment
**DTF**    Domain Task Force
**EACL**    Environment Adaptation Class Library
**ESIOP**    Environment Specific Inter-ORB Protocol
**ESP**    Environment of Service Provisioning
**ETSI**    European Telecommunications Standards Institute
**FCAPS**    Fault, Configuration, Accounting, Performance and Security
**FE**    Functional Entity
**FEC**    Forwarding Equivalence Classes
**FGP**    First Generation Prototype
**FIPA**    Foundation for Intelligent Physical Agents
**FSM**    Finite State Machine
**FTP**    File Transfer Protocol
**GDMO**    Guidelines for the Definition of Managed Objects
**GIF**    Graphics Interchange Format
**GIOP**    General Inter-ORB Protocol
**GPRS**    General Packet Radio Service
**GSM**    Global System for Mobile Communication
**GUI**    Graphical User Interface
**HAP**    Home Agent Platform

**HMP**     Home Mobility Provider
**HSCSD**     High Speed Circuit Switching Data
**HTML**     Hyper Text Markup Language
**HTTP**     Hyper Text Transfer Protocol
**IA**     Infrastructure Agent
**IA**     Intelligent Agent
**IAM**     Initial Address Acknowledge
**IBN**     Intelligent Broadband Network
**ICMP**     Internet Control Message Protocol
**IDL**     Interface Definition Language
**IETF**     Internet Engineering Task Force
**IGMP**     Internet Group Management Protocol
**IIOP**     Internet Inter-ORB Protocol
**IMR**     Interactive Multimedia Retrieval
**IMT-2000**     International Mobile Telecommunications - 2000
**IN**     Intelligent Network
**INAP**     Intelligent Network Application Protocol
**INCM**     Intelligent Network Conceptual Model
**IN-SSM**     Intelligent Network Switching State Model
**IP**     Internet Protocol
**IP**     Intelligent Peripheral (in the context of IN)
**IPC**     Interprocess Communication
**IPv4**     Internet Protocol version 4
**ISDN**     Integrated Services Digital Network
**ISO**     International Standardization Organization
**ISP**     Internet Service Provider
**IT**     Information Technology
**ITU-T**     International Telecommunication Union – Telecommunications sector
**JVM**     Java Virtual Machine
**KIF**     Knowledge Interchange Format
**KQML**     Knowledge Query and Manipulation Language
**LAN**     Local Area Network
**MA**     Mobile Agent
**MAC**     Medium Access Control
**MAC**     Message Authentication Code
**MAN**     Metropolitan Area Networks
**MAP**     Mobile Agent Platform
**MAS**     Multi Agent Systems
**MASIF**     Mobile Agent System Interoperability Facility
**MAT**     Mobile Agent Technology
**M-BCUSM**     Mobility Basic Call Unrelated State Model
**MbD**     Management by Delegation
**M-CUSF**     Mobility Call Unrelated Service Function
**MHEG**     Multimedia Hypermedia Experts Group
**MIB**     Management Information Base
**MKH**     Mobility Key Handler
**MM**     Mobility Management

**MO**     Managed Object
**MOA**    Managed Object Agent
**MPEG**    Moving Picture Experts Group
**MPLS**    Multi Protocol Level Switching
**MPOA**    Multi Protocol Over ATM
**MSA**    Messaging Service Agent
**M-SCF**    Mobility Service Control Function
**M-SDF**    Mobility Service Data Function
**MSS**    Messaging Service
**NC**    Network Computer
**NE**    Network Element
**NFD**    Notification Forwarding Domain
**NNI**    Network Node Interface
**NT-2**    Network Termination 2
**OA**    Object Adapter
**OAM&P**    Operations Administration Management and Provision
**ODP**    Open Distributed Processing
**OLE**    Object Linking and Embedding
**OMA**    Object Management Architecture
**OMG**    Object Management Group
**OMT**    Object Modeling Technique
**OO**    Object Oriented
**OOA**    Object Oriented Analysis
**OOADA**    Object Oriented Analysis and Design with Applications
**OOD**    Object Oriented Design
**OOP**    Object Oriented Programming
**OOSE**    Object Oriented Software Engineering
**ORB**    Object Request Broker
**ORBOS**    Object Request Broker and Object Services
**ORPC**    Object Remote Procedure Call
**OS**    Operating System
**OSF**    Open Software Foundation
**OSI**    Open Systems Interconnections
**PARC**    Palo Alto Research Center
**PC**    Program Counter
**PDN**    Public Data Network
**PE**    Physical Element
**PHB**    Per Hob Behavior
**PIC**    Points In Call
**PLMN**    Public Land Mobile Network
**PoA**    Point of Attachment
**POA**    Portable Object Adapter
**POTS**    Plain Old Telephone Service
**PRI**    Primary Rate Interface
**PSE**    Packet Switching Exchanges
**PSIG**    Platform Special Interest Group
**PSTN**    Public Switched Telephone Network

**QoS**   Quality of Service
**RA**    Rule Agent
**RB**    Roaming Broker
**REV**   Remote Evaluation
**RFI**   Request For Information
**RFP**   Request For Proposal
**RISC**  Reduced Instruction Set Computer
**RLA**   Resource Logic Agent
**RLP**   Resource Logic Program
**RMI**   Remote Method Invocation
**RPC**   Remote Procedure Call
**RSVP**  Resource Reservation Protocol
**SA**    Service Agent
**SCE**   Service Creation Environment
**SCEF**  Service Creation Environment Function
**SCF**   Service Control Function
**SCP**   Service Control Point
**SCS**   Service Creation System
**SCUAF** Service Control User Agent Function
**SDF**   Service Data Function
**SDH**   Synchronous Digital Hierarchy
**SDL**   Specification and Description Language
**SEE**   Service Execution Environment
**SEN**   Service Execution Node
**SGP**   Second Generation Prototype
**SIB**   Service Independent Building Block
**SIF**   Service Intelligence Framework
**SIG**   Special Interest Group
**SL**    Service Logic
**SLA**   Service Logic Agent
**SLM**   Service Logic Manager
**SLP**   Service Logic Program
**SMAF**  Service Management Access Function
**SMC**   Service Management Controller
**SMF**   Service Management Function
**SMS**   Service Management System
**SMTP**  Simple Mail Transfer Protocol
**SN**    Service Node
**SNMP**  Serving Network Mobility Provider
**SP**    Service Provider
**SPD**   Service Provisioning Domain
**SRF**   Specialized Resource Function
**SS&CP** Service Switching and Control Point
**SS7**   Signaling System No. 7
**SSCOP** Service Specific Connection Oriented Protocol
**SSF**   Service Switching Function
**SSF-AM** Service Switching Function Access Manager

**SSL**    Secure Socket Layer
**SSM**    Switching State Model
**SSP**    Service Switching Point
**STB**    Set Top Box
**TCAP**    Transaction Capabilities Application Protocol
**Tcl**    Tool command language
**TCP**    Transmission Control Protocol
**TDM**    Time Division Multiplexing
**TDMA**    Time Division Multiple Access
**TE**    Terminal Equipment
**TE-1**    Terminal Equipment 1
**TE-2**    Terminal Equipment 2
**TINA-C**    Telecommunications Information Networking Architecture Consortium
**TMN**    Telecommunications Management Network
**TMTI**    Temporary Mobile Terminal Identity
**TMUI**    Temporary Mobile User Identity
**UA**    User Assistant
**UCSD**    University of California at San Diego
**UDP**    User Datagram Protocol
**UIM**    User Interaction Manager
**UML**    Unified Modeling Language
**UMS**    User Mobility Services
**UMTS**    Universal Mobile Telecommunications Service
**UNI**    User to Network Interface
**USI**    User Service Interaction
**VBR**    Variable Bit Rate
**VC**    Virtual Circuit
**VCI**    Virtual Circuit Identifier
**VERA**    Environment Monitoring Agent
**VHE**    Virtual Home Environment
**VLSI**    Very Large Scale Integration
**VM**    Virtual Machine
**VoIP**    Voice Over IP (Internet Protocol)
**VP**    Virtual Path
**VPI**    Virtual Path Identifier
**VS**    Video Server
**WAN**    Wide Area Networks
**WWW**    World Wide Web

# Part I

The Need for Advanced
Software Technologies in
Telecommunication Networks

# 1

# Networks and Telecommunications Software Evolution

## 1.1   Introduction

Let's accept for a moment that the use of the word 'information' in the context of the next few sentences is unambiguously defined. The world abounds with information sources and potential users of that information. To maintain the analogy, one could term the latter 'information sinks'. Technology is put to use to exploit physical phenomena in a way that can allow information to be transferred from sources to sinks. Inherent in that approach is a necessary step of encoding information into a more tangible physical quality, one that observes the laws of physics and that, at least at a certain level, exhibits deterministic behavior. Nineteenth century telegraphy used electric signals propagating through cooper wires to transfer a Morse-encoded representation of a text. There is nothing inherently different between a system of telegraphy stations and an equivalent one of smoke-signal stations. Yet, even using as examples two outdated forms of communication, we have included an additional parameter that has extended our model of information sources and sinks. Implicitly, the assumption in both examples has been that the volume of information to be transferred and the number of potential users of it, render dedicated, *ad hoc* arrangements unpractical. Instead, the volume of information and the frequency with which exchanges of it occur warrants a permanent infrastructure. This infrastructure will require a significant investment to be put into place, and will have associated operational, management and upkeep costs. Its scope and the frequency with which it will be utilized, however, will allow it to exploit economies of scale in all these functions.

Networks can therefore be defined as systematic arrangements of information processing and conveying elements put in place with the purpose of providing a permanent infrastructure for information needs. Permanent, at least, with respect to typical information exchange's time scales. Using this simple model of information sources and sinks, and viewing networks as the serving infrastructure, one can attempt to reason about the

*Object Oriented Software Technologies in Telecommunications*, Edited by I. Venieris, F. Zizza and T. Magedanz
© 2000 John Wiley & Sons Ltd

interdependencies between these elements of the model. For instance, what constitutes information? What are the mechanisms and technology used at the source to transform a form of information into a representation that would allow it to employ the laid network and be transformed back to the same form at the sink? What properties can be ascribed to information, and to what extent will the design of the network infrastructure be dependant on them or reflect them? What kind of economic considerations underpin the further design and elaboration of a network scheme, and what, if any, are the trade-offs back to any ascribed information properties?

Not surprisingly, it turns out that asking these questions is as far as one can go into reasoning about these issues without any specific technological paradigms in mind. The model, similarly, cannot be further extended or refined without impairing its inclusiveness in scope. So one, reluctantly, has to give up meditating in the abstract realm, and concretize at least the less developed terms. It is necessary to replace generic notions and model elements with real world technologies and concepts, a certain degree of familiarity with which is assumed for the remainder of this chapter.

More specifically, for the purposes of this book, we will consider information as being either voice or moving/still pictures or data. It is easy to identify what the sources and sinks of information are in any case. The telephone network is used to transfer voice. We will use the term 'telecommunication networks' to refer in general to high-throughput networks that are used to convey stream-like media such as audio or video. Networks used to transfer information taking the form of computer data will be referred to as data networks or computer networks. It will be shown that the questions asked above can be asked again with more success when having a specific network and types of information in mind. In fact, we will for the most part disregard the transformation or encoding issue, and concentrate more on the mechanisms that direct the flow of information through the network. Also, since the properties of telephone and telecommunication networks are remarkably similar from a qualitative point of view, we will use the latter term as encompassing the first, and will otherwise treat them for the most part alike.

## 1.2    A Unifying Perspective of Networking Technologies

The telephone network is the world's biggest information network. It is also one that has evolved significantly since its original conception, in spite of the fact that the basic service it offers has remained unchanged. In this way its history has demonstrated the ways in which economic considerations and technical innovation shape networks not in how their users perceive them, but on the way they operate inside.

Returning to the questions mentioned above, one would first ask what are the properties of voice. Human voice is characterized by a natural, high redundancy, at least for the purposes of an adequately meaningful conversation. Secondly, a typical phone call lasts an amount of time which can appropriately be modeled by a random variable. Using accumulated call duration data and regression analysis techniques, one can estimate appropriate distributions and parameters. Similar techniques can apply when seeking values for such variables as mean elapsed time between calls or, perhaps more informatively, an estimation of the stochastic process that generates these calls. Last but not least, there are a lot of behavioral aspects which are typical of a pair of telephone users that played a role in the design of the telephone network, and which have made their way into its architecture and protocols over the years. They revolve around the important issue of the quality of service a user expects or is accustomed to expect. Returning to the abstract model of a set of information sources and sinks that exchange information by means of a network, one can represent the information arriving at a sink as a distorted version of the information stemming from the source. That is, the network has altered the representation of the information flowing through it and, in so doing, the information itself. For instance, in a typical telephone conversation the electric signals arriving at the receiver's end represent, in general, a distorted, delayed and attenuated version of the waveform that left the transmitter's end. Since human voice can be comprehensible even when significantly distorted, users of the telephone network do not seriously object to experiencing a less than optimal reproduction of their peer's voice. What is less tolerable, however, is the possibility that a phone conversation will be abruptly stopped. Users can accept having their calls rejected by the network in case of congestion, but not after communication with the called party has begun.

The intolerance to a degrading quality of a conversation points to a connection oriented service implementation on behalf of the network. Connection oriented services are those where, prior to initiation, resources are committed along every network link connecting the sender to the receiver. Also, there is a lot of state information associated with the call, and this information is maintained for the duration of the call, to be discarded only after its completion. This step guarantees that either a call will be blocked at the beginning, or that once resources are committed, the user can expect a relatively stable quality of service. Actually, the term 'connection oriented service' can also be applied to services satisfying only the second, less demanding criterion. That is, a service can be termed 'connection oriented' simply if it maintains some sort of state information at either endpoint even without committing resources internally to the network. Symmetrically, a service will be termed 'connectionless' if it is not maintaining any state information for the duration of an information exchange, but simply treats each individual segment of information to be conveyed across the network as distinct from the others.

The technological means by which the first telephone networks were implemented (electromechanical equipment relaying a physical, electric current path) mandated an implementation that was inherently connection oriented. This sort of switching implementation is termed 'circuit switching'. In it, especially in its first implementation for the telephone network, each connection results in a physical communication channel through the network from the calling to the called subscriber equipment. This connection is used exclusively by the two subscribers for the duration of the call, providing a fixed bandwidth channel. This line is dedicated, regardless of whether or not communication actually takes place, and no further processing or routing calculations are needed.

The physical characteristics of the early telephone network and the technological status at that time allowed few alternatives. If one considers more advanced technological equipment such as that which is typically available today, there is nothing precluding a connection-oriented service from using at a certain level non circuit-switching techniques.

In traditional telecommunication networks, where analog transmission was used, phone conversations were multiplexed on a single wire by means of *Frequency Division Multiplexing* (FDM) or static *Time Division Multiplexing* (TDM) techniques. Otherwise, however, individual connections perceived a physical medium devoted to them, end-to-end. The general argument against circuit switching techniques is that a significant percentage of communication links is under-utilized when there is no information to be transmitted because other connections cannot employ this underused capacity. This is because resource commitment is static in nature. One has to note, however, that the relatively long duration of a typical phone call and the continuous use of the medium by the users during that time make this potential shortcoming less noticeable. In the case of a bursty and relatively discontinuous flow of information, however, this deficiency becomes more evident.

It was not until advancements in solid state technology and *Very Large Scale Integration* (VLSI) design dropped chip costs and afforded significant processing power to network designers that digitization and packet switching became possible. Digitization of voice, or in general, audiovisual information is a processing power consuming operation that requires sampling at an adequately high rate followed by quantization and encoding. At this point a video or audio signal has been transformed into a series of bits. Usually, a further step of compression is used to exploit natural redundancies in both audio and video signals, whereby the bit stream is reduced in length without or with negligible loss of actual user-perceived information. With digitization, the bit stream produced can be segmented into packets that flow through the network in digital form. As before, this is ultimately accomplished by the propagation of an electromagnetic waveform through a physical medium. However, since the waveform, being now interpreted as a digital signal, has a much more structured and restricted form, successive repeaters can spot noise and accurately reproduce the transmitted signal, resulting in clearer reception. Moreover, when the signal reaches a bit stream format, it is possible to introduce controlled redundancy in the form of checksums that can further help detect or even correct transmission errors.

It can be seen that with digitization the number and kind of transformations applicable to information submitted to the network for delivery goes beyond the simpler transformations that are possible in analog networks. As digitization enables precise reproduction of a signal, and facilitates storing it or otherwise processing it, more elaborate schemes can be defined. These schemes essentially substitute processing power for physical medium resources, and better exploit the available infrastructure. Without the historical advances in chip technology, those substitutions would be uneconomical. In contrast, there is no meaningful processing to apply to an analog signal (except perhaps shifting it to another frequency band), nor is it possible to view and examine it in a 'frozen' state. As systems become more complex a black box approach is appropriate to allow designers and implementers to tackle complexity and concentrate only on the pertinent issues. Where a certain black-box strategy is observed by more than two peers, layering arises. While layering is a pretty generic concept, it did not arise in the sense that is currently employed before the advent of computer networks or the digitization of telecommunication ones, since traditional engineering disciplines usually had little need of defining intelligent behavior at different conceptual levels.

In a typical telecommunication network, an audiovisual signal is provided to the first black box, whose output is then fed to a second, and so on, until it takes the form of a waveform on a physical medium. This waveform traverses the network, and at certain nodes is momentarily elevated to an intermediate level before again being transmitted as a waveform. At the destination, the reverse procedure is followed this time all the way up to the actual audiovisual information that entered the first black box on the transmitter's side. Since this process defines an hierarchy of black boxes, and that hierarchy has the same structure at both connection ends, as well as in intermediate nodes where applicable, we can refer to a layered organization.

The *Open Systems Interconnection* (OSI) reference model was designed with the goal of defining such a layering for telecommunication networks. It is shown in Figure 1.1. It defines seven layers in total. A detailed presentation of the OSI reference model would require many pages, and essentially almost every introductory book on networking provides one. We will only make some quick notes, mainly to introduce terms that will be used in the following sections. The physical layer is the black box responsible for transmitting bits in a synchronous manner with a non-zero probability of success over a physical medium between two adjacent network nodes. The data link layer is responsible for converting the unreliable synchronous bit pipe offered by the physical layer into a reliable bit link over which packets can be sent asynchronously. The network layer is responsible for routing packets from source to destination, and is conceptually the most complex, since in this layer all intermediate nodes must work in concert. The transport layer is responsible for breaking data provided by the application into packets to be submitted to the network layer black box. Finally, the session, presentation and application layers are responsible for higher layer functions such as access rights, encryption and necessary data coding transformations between different formats. Each layer observes two things: the interface provided to it by the immediately lower layer – which it views as a black box – is the first. Each layer receives from its lower layer information that has passed through a number of nodes before reaching it, but conceptually regards it as having being sent by its peer layer on the other end point of a connection. The same applies to the information it itself provides to its lower layer. It views it as being directly directed to its peer layer, although it will undergo many transformations before actually arriving there. The term 'protocol' is used to describe the set of conventions and rules observed by the two peer layers in this communication.

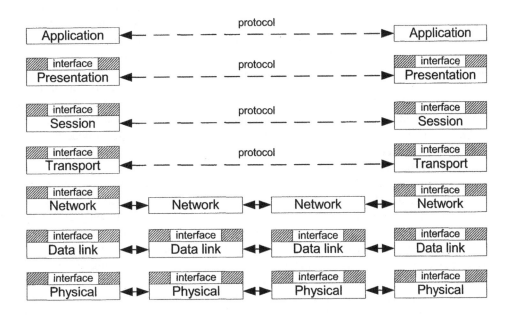

**Figure 1.1**  The OSI reference model

Returning to our discussion of the circuit switching technique, we note that the basic advantage of the so-called packet switching techniques over circuit switching ones is that the link under-utilization problem can be solved. This is because each link can be fully utilized wherever there is any offered traffic available. The segmentation of a stream like information source into discrete identifiable packets enables the network to treat each packet individually if it wishes to do so, and thus allows wider latitude in its routing decisions. This is the basic concept of packet switching. As a result, the network infrastructure can be better exploited than in the case of static frequency bands or time division slots allocated throughout the network between the two endpoints of a connection.

Packet switching, however, also introduces an additional parameter, which is the timing relationship between the transmission and reception of individual packets. For stream like media, it is necessary that this timing relationship is maintained or otherwise the voice or video signal reproduced at the receiving end will be incomprehensible. A constant delay does not pose any problem. However, variation in this delay, known as jitter, does. Jitter is non-existent in circuit switching techniques using either FDM or TDM, since the propagation speed of the electromagnetic wave is constant, and so is the distance traveled by each piece of information. Packets are subject to processing delays, however, which can vary. Also, if the network is allowed to dispatch individual packets as it sees fit, they may end up taking different routes, passing from different links, and be subject to different transmission times. As a result jitter, may increase unexpectedly. Since most types of information to be conveyed by telecommunication networks are jitter sensitive, a more restrictive form of packet switching, known as virtual circuit switching, is often employed.

Virtual circuit switching is a packet-based technique whereby, prior to the initiation of a connection, a path is specified across the network and optionally resources for it are committed. All packets belonging to the same connection follow this path from source to destination, and are thus subject to roughly the same delays. In this way jitter is minimized. Virtual circuit switching is similar to 'real' circuit switching in that a path is fixed for all packets, but it is different in that this path exists only at the packet level and not at the physical level. That is, the physical medium's transmission capacity is not segmented by means of allocating fixed frequency bands or time division slots on every link. Prior to digitization in the telephone network, circuit switching meant that an electromagnetic waveform would transcend it almost uninterrupted. Virtual circuit switching is quite different. Rather, adjacent nodes in the network simply collaborate to provide the same routing for all packets of a connection. Accordingly, there are a lot of messages that need to be exchanged between network nodes along the path from the calling to the called party before establishing the call. Since considerable resources are required for the network to set up a virtual circuit, a lot of state information has to be maintained, and so circuit switching is almost entirely used for connection oriented services. However, nothing precludes a virtual circuit from being set for transferring only a subset of packets belonging to a connection, and having a second, third, and so on set for transferring subsequent subsets. Even more meaningful can be the case where 'real' circuit switching is used in a connectionless manner instead of packet switching. Circuit switching is costly since the time to set up the call can be significant. Once the connection is in place, however, signals propagate with increased speed and there are no further delays (no more routing decisions to be made). In contrast, a packet switched network has no set-up delays, but packets delay for processing on each intermediate node and are transmitted as many times as are the links in between. Depending on the numeric values, it can be opportune to set up a circuit even for a few bytes, transmit them and then tear the connection up, setting a new one once more bytes are available. This concept is referred to as 'fast connect' circuit switching.

Attempting to define a taxonomy of networks is error prone and will certainly fail in one respect or another. Nevertheless, using a limited set of properties as distinguishing characteristics, a scheme can be produced. Figure 1.2 depicts such a scheme, where the set of properties consists of the transmission technique (analog or digital), the connection characteristics of the 'basic' service (connection oriented or connectionless), and the switching technique (circuit versus packet switching). The scheme is not normative, but is intended as an intuitive (although rough) way to depict some of the concepts developed in this chapter.

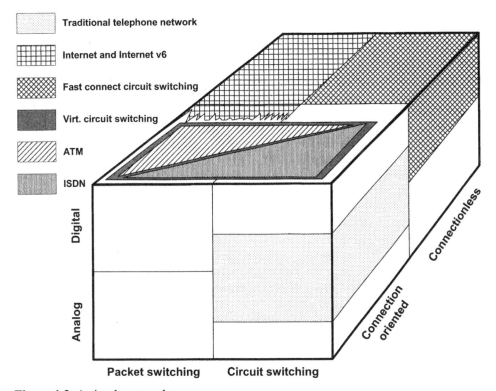

**Figure 1.2** A simple network taxonomy

Nowadays, the telephone network is digitized in its core while the local loop that connects individual subscribers to the local exchanges remains analog. So, digitization and quantization take place at the first local exchange, packets are transmitted to the destination's local exchange in a (virtual) circuit switching manner, and there the digital signal is reconstructed and converted back to analog form to cover the 'last mile' to the receiver's telephone appliance. In spite of the fact that the telephone network cannot be used as it is to provide digital connectivity end-to-end, modems can be employed to allow for digital communication to take place over the analog parts of it. Since the core of the telephone network is already digital, it is mainly the local loop and the switching technology that differentiates modern telecommunication networks from typical computer communication networks like the Internet. Internet technologies and their evolution are examined in more detail in Section 1.4, but for the moment, as Figure 1.2 shows, Internet finds its place in this simple taxonomy as a digital, connectionless, packet switching network. While the Internet suite of protocols contains higher level connection-oriented protocols, its routing protocols are connectionless in the sense that they do not get into the process of committing resources before establishing a connection, nor of treating similarly packets belonging to the same connection. For the purposes of routing, the Internet treats packets belonging to the same session as distinct from each other. In fact, at the network layer, it is not able to tell if two packets belong to the same application, since it only perceives node addresses. As such, packets may be subject to different routing decisions. Each router examines each packet individually and decides on which link to forward it based only on its present routing table condition (which may change in the duration of a connection). Intermediate routers do not collaborate between them to provide a virtual circuit for all packets belonging to a connection. In fact, routers (as in general layer 3 entities in every network) do cooperate as part of the routing policy decisions they have to take, but this collaboration is unrelated to the serving of a particular information exchange. It happens on different time scales, on a different level of abstraction, and using routing protocols with which the user of the network has no means to interact. Furthermore, this process is not triggered by any request on behalf of the user to set up a connection. This is in contrast with what happens in virtual circuit switching networks (see Section 1.3), where collaboration between intermediate nodes is triggered by the user's request to set up a connection and in response to the need to provide uniform routing to all packets belonging to that connection.

Protocols proposed to complement the current Internet protocols such as the *Resource Reservation Protocol* (RSVP) will probably be employed in future versions of the Internet (Internet v6, also known as 'next generation Internet'), and will provide some sort of a connection oriented routing service. Currently, the Internet provides a connection oriented transport layer service through the *Transmission Control Protocol* (TCP). However, at the network layer, the *Internet Protocol* (IP) is connectionless, and that's why most of the Internet falls in the connectionless area of the taxonomy. However, since both RSVP as well as similar technologies like *differentiated services* are associated with the network layer (the IP protocol), even the routing service itself will become more connection oriented in future versions of the Internet.

Rough though the taxonomy depicted in Figure 1.2 may be, it is still revealing that telecommunication networks and data networks are converging. In Chapter 2 we will make the case that this convergence is also manifested in the service engineering realm. It will be further shown that, fostered by this convergence, methodologies are also converging, and that in the software field, paradigms of development and associated practices can more easily be diffused between telecommunication and data networks. Since the latter have from their inception been digital and have relied heavily on software due to their greater processing needs, this diffusion is not actually symmetric. It is rather that practices and procedures cultivated in data networks are beginning to make their way into telecommunications as the latter become more similar to data networks. This trend has gathered pace over the last few years, and will continue to do so as convergence becomes more complete.

Modems are meant as a short term solution. The future telecommunication networks aim for digital connectivity end-to-end. Since the new digital local loop will offer greater bandwidth than is currently available in a typical phone line (taking into account the effect of filters restricting that bandwidth), more services will become possible. In addition to the transfer of voice at a higher quality, it will also be possible to transfer video. The set of services offered by the telephone network will change significantly, as will patterns of traffic and dimensioning. Instead of a network infrastructure to convey voice, the new 'telephone' network will also convey video, and it will also be possible to convey non-stream media like computer data. The vision is that instead of separate networks for each type of information, a single network will accept and transfer all. That vision took shape in the specification for an *Integrated Services Digital Network* (ISDN). This was soon followed by the *Broadband Integrated Services Digital Network* (B-ISDN) that sought to exploit advances in fiber optic technology and a new mode of switching called *Asynchronous Transfer Mode* (ATM). The following sections describe these and other evolutions in the telecommunication networks arena, and the way they have affected service engineering and management operations. Section 1.4 discusses the Internet, which is the prevailing data communication network.

## 1.3    Telecommunication Networks Technologies

Network evolution during the last few decades has been impressive. In the traditional *Public Switched Telephone Network* (PSTN) the only service offered was classical, two-way voice conversation. Modern broadband and mobile services networks are able to support complicated and widely diffused services. The networks evolution has been characterized by a transition from single service networks to integrated services ones. The fast and efficient provision of new services in response to varying market needs and orientation is another matter of major significance. Apart from the network infrastructure itself, the corresponding software has also significantly evolved. This evolution is reflected in miscellaneous yet interrelated areas such as signaling (control) software, the network management software and the application of distributed processing paradigms.

Contemporary public networks are highly specialized. PSTN as we noted basically offers the *Plain Old Telephone Service* (POTS). Digitization and use of Intelligent Network concepts has allowed it to offer a few more complementary services. Intelligent Networks are treated in more detail in Part III, but can basically be thought of as a control network overlaid on top of PSTN that is transparent to its switching procedures, and has thus minimal impact on existing equipment. In this way, variation of the basic phone service with respect, for instance, to charging can be efficiently implemented. Even IN-augmented PSTN, however, fails to meet the service demands of the future.

On the other hand, while the use of modems can allow exploitation of PSTN infrastructure for data communication purposes, there are usually *Public Data Networks* (PDN) for that purpose. The two main types of Public Data Networks are again circuit switched, based on the X.21 standard, and packet switched, based on the X.25 protocol. Moreover, the Internet has arisen as a ubiquitous low-cost data network for the private domain. It can also be used for voice connection but, at least for the time being, with a markedly lower quality than that offered by PSTN. While PDNs can share much of a

PSTN's infrastructure in terms of cable equipment or wireless links, from a service point of view as well as administratively, they represent separate networks.

The telecommunication's landscape described above is specialized, and this specialization in services is inefficient. Management, operating procedures and functions are reproduced for each network, as are costs. A single integrated services network would be more appropriate. *Integrated Service Digital Network* (ISDN) and *Broadband ISDN* (B-ISDN) have emerged to cater for the high-cost, high-throughput market segment. In contrast to packet switched networks like the Internet, B-ISDN is based on Asynchronous Transfer Mode technology, which makes it a virtual circuit switching network. There are economic considerations underpinning any preference for packet based networks over circuit switched ones, or the other way round. Packet switching, as we have shown, is able to better utilize the available network capacity at the expense of processing equipment in network switches. During the last few decades the cost of both communications links and computer hardware and software has been declining. A point was reached were switches became relatively cheaper than lines. At that point, packet switching, became cost-effective as lines are shared by multiple connections at the cost of more routing calculations by the switches. The preference for packet switching techniques (in which we also count virtual circuit switching) over circuit switching techniques is likely to continue as it is bolstered by falling computing prices. Even between packet based networks, the connection oriented or connectionless nature can also account for differences in ease of deployment, associated administrative and management costs and flexibility. Nowadays, connectionless, packet switching Internet is a cheaper and more easily deployed alternative than ATM, although it cannot compete with it in terms of throughput and reliability.

## 1.3.1    Historical evolution of telecommunications networks

The PSTN was the first and most widespread telecommunications network. PSTN is the global collection of interconnected voice-oriented public telephone networks, both commercial and government-owned. It aggregates national and regional circuit switching telephone networks, operated by the various telephone companies and administrations around the world. It is also referred to as the Plain Old Telephone System. The basic service offered by PSTN is to provide voice-grade, circuit based connections on demand. The connection is switched by virtue of the called party's telephone number dialed by the originator of a call. A permanent circuit can also be established over a switched circuit system by simply bypassing the switching function. In general, such a circuit may be useful for industrial or commercial customers since it is usually charged a flat price.

The PSTN's first network nodes were human-operated analogue circuit switching systems. Its first evolutionary step was to replace human operators with electromechanical switches. Phone appliances would then communicate the intended destination to these switches by producing appropriate analog waveforms. Replacing these switches with digital processing equipment came next, at which point it became possible for nodes internal to the network to collaborate for routing by using 'proper' protocols. For signaling purposes between these nodes, a separate signaling network was used (a common such network is the *Signaling System 7* (SS7) network). Currently, in most countries the telephone network has been almost completely digitized, except for the final connection from the local exchange to the subscriber. Since the circuitry used in these 'local loops' accounts for many times the digitized circuitry interconnecting local exchanges between them and with other nodes higher in the hierarchy, this process may take a long time to come to completion. Therefore, currently, the signal coming out of the phone set is analogue. It is usually transmitted over a twisted pair cable. At the local switch the analogue signal is usually digitized, using 8000 samples per second and 8 bits per sample, yielding a 64 Kbps Data Stream (DS0). Several such data streams are then combined to form a new one. Twenty four such channels are combined into a T1 (in the US), while 30 channels (in Europe) form an E1 line. These signals can be further combined into larger chunks for transmission over high-bandwidth core trunks. At the receiving end the channels are split, and the digital signals are converted back to analog and delivered to the received phone.

Initially the synchronization of signals was not necessary. It was the baseband signal that was transferred and only one channel was available per line, requiring large quantities of wiring to meet the growing communication demand. Nowadays, multiplexing techniques permit the simultaneous transfer of multiple data flows over the same line. PSTN is based on Time Division Multiplexing. The multiplexing and demultiplexing of voice traffic works well in a fairly static, voice-oriented environment. The network offers high reliability, low-latency and jitter, scalable call routing through the SS7 infrastructure and value-added voice services such as voice messaging and caller ID recognition. Dramatic drops in the price of digital electronics made not only the switches less costly, but it also made high-speed long-distance digital interconnection less costly than dedicated lines. Fiber-optic lines connect many distant points in the PSTN with data rates of 612 Mbps. Thus, very high bit rate digital circuit channels can be sliced out of the available connection for resale to digital services subscribers. In effect, the PSTN merely dedicates a percentage of the bit stream capacity that would otherwise have served a number of telephone conversations to transport the digital data. Of course the user in such a case must conform to certain digital data speeds and formats.

To provide digital connections end-to-end between two users of a telecommunication network, the International Telecommunications Union (ITU) produced the ISDN set of standards for digital transmission over ordinary telephone copper wire as well as over other media. ISDN in concept is the integration of both analogue or voice data together with digital data over the same network. Basically, the term ISDN signifies a telecommunications service package, consisting of digital facilities end-to-end. Digital circuits are typically characterized by very low error rates and high reliability. ISDN can support voice, video and data transmissions.

According to ISDN, individual channels can transmit at speeds up to 64 Kbps, and these channels are combined to support higher speeds. Each bearer channel (B channel) transmits at 64 Kbps. A data channel (D channel) operates at 16 Kbps or 64 Kbps, depending on the service type. There are two basic types of ISDN service: *Basic Rate Interface* (BRI) and *Primary Rate Interface* (PRI). The BRI consists of two 64 Kbps B channels and one 16 Kbps D channel that can handle signaling and potentially data, offering a total rate of 144 Kbps. This service is commonly referred to as 2B+D service. It is intended to meet the basic needs of most residential users. For instance in BRI, the first B channel could be used for phone conversation, the second for an extra voice conversation or data communication purposes, whereas the D channel could be used either for signaling or for telemetry applications (e.g. alarms). The PRI is intended for users with greater capacity requirements, consisting of 23 B channels (30 in Europe) plus one 64 Kbps D channel for a total of 1536 Kbps (1984 Kbps respectively in Europe). In PRI, the D channel is used exclusively for signaling and network functions regarding the B channels. Customers must be within a certain distance from the local ISDN switch (approx. 3.4 miles); beyond that expensive repeater devices are required, or ISDN service may not be available at all.

ISDN allows multiple digital channels to operate simultaneously through the same regular phone wire. The same physical wiring is used to transmit a digital signal across the line rather than an analog waveform. This scheme permits a much higher data transfer rate than analog lines. In the traditional PSTN the number of required phone lines is equal to the number of Terminal Equipment (TE) devices that operate at the same time. ISDN enables the combination of many different digital data sources and the routing of the relevant information to the proper destination. Since the line is digital, it is easier to keep the noise and interference out while combining these signals. ISDN technically refers to a specific set of digital services provided through a single, standard interface. Without ISDN, distinct interfaces would be required instead.

ISDN is the first case where Out-Of-Band signaling is used. The ring voltage signal to the *Customer Premises Equipment* (CPE) is replaced by a digital packet sent on a separate channel. The Out-of-Band signal does not disturb established connections, and call setup time is considerably reduced. The signaling also indicates who is calling and the type of call. Available ISDN phone equipment is then capable of making intelligent decisions on how to direct the call.

Devices most commonly expect either a U interface connection or an S/T interface connection. The U interface is a two-wire (single pair) interface from the phone switch. It supports full-duplex data transfer over a single pair of wires, therefore only a single device can be connected to a U interface. This device is called a Network Termination 1 (NT-1). The NT-1 is a relatively simple device that converts the 2-wire U interface into the 4-wire S/T interface. The S/T interface supports multiple devices (up to seven devices can be placed on the S/T bus) because, while it is still a full-duplex interface, there is now a pair of wires to receive data, and another to transmit it. When devices have NT-1 built into their design they are less expensive and easier to install, but flexibility is reduced as additional devices cannot be connected.

ISDN devices must go through a Network Termination 2 (NT-2) device, which converts the T interface into the S interface. Virtually all ISDN devices include an NT-2 in their design. The NT-2 communicates with terminal equipment. Devices that connect to the S/T (or S) interface include ISDN capable telephones and fax machines, video teleconferencing equipment, bridge/routers, and terminal adapters. All devices that are designed for ISDN are designated Terminal Equipment 1 (TE1). All other communication devices that are not ISDN capable, but have a POTS telephone interface (also called the R interface), including ordinary analog telephones, fax machines, and modems, are designated Terminal Equipment 2 (TE2). A Terminal Adapter (TA) is used to connect a TE2 to an ISDN S/T bus. Figure 1.3 depicts that arrangement.

ISDN offers notable price performance advantages for certain types of applications. It offers traffic rate and transmission reliability that can significantly enhance the connections between a remote office to a *Local Area Network* (LAN), as compared with technologies based on modems over analog lines. It can also support a videoconference in the context of either business environments or residential use. The fact that ISDN connections can be used for a specific time frame renders them a cost-effective solution for interconnecting LANs that do not have to be constantly connected. As the cost of leased lines is often prohibitive, ISDN can satisfy the growing needs of connecting remote offices to the corporate LAN backbone.

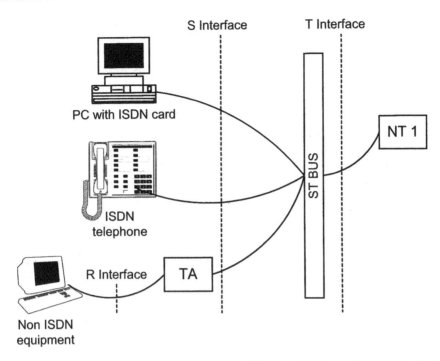

**Figure 1.3** Arrangement of NT1, ISDN, non-ISDN and TA equipment and with respect to S and T interfaces

While ISDN was designed with mostly stream like media in mind, the rapid growth of distributed data processing and the Internet changed the situation. The bursty characteristics and bandwidth demands of data transmission were not well suited to the circuit switched nature of the ISDN. ISDN's fixed bandwidth architecture was proven to be inefficient. A new solution was required. Public data networks often use packet switching techniques to accommodate the bursty characteristics of data applications more efficiently than the ISDN's fixed 64 Kbps architecture.

In packet switched networks all data to be transmitted is first assembled into packets by the source *Data Terminal Equipment* (DTE). These packets include both the source and the destination DTE network addresses. Using appropriate routing tables, the packets are transferred via intermediate *Packet Switching Exchanges* (PSE) to the destination DTE. Packets are sent from the source DTE to the local PSE. On receipt of each packet, the PSE inspects the destination address contained in the packet. Based on a routing directory specifying the outgoing links to be used for each network address, it forwards the packet on the appropriate link at the maximum available bit rate. As each packet is received at each intermediate PSE along the route, it is forwarded on the appropriate link interspersed with other packets with which it shares that link. At the destination PSE the packet is finally delivered to the destination DTE.

To prevent unpredictably long delays and to ensure that the network has a reliably fast transit time, a maximum length is allowed for each packet. A message submitted to the transport layer within the DTE may first have to be divided by the transport protocol entity into a number of packet units before transmission. In turn, they will be reassembled into a single message at the destination DTE.

In packet switching networks error and flow control procedures are applied on each link by the network PSEs. Two types of services are normally supported: datagrams and virtual circuits. In the datagram service, each packet entering the network is treated as a self-contained entity with no relationship to other packets. This former case is used for the transfer of short, single packet messages.

For a message containing multiple packets a virtual circuit service is normally selected. Prior to the information transfer phase a Virtual Circuit (VC) is established. A virtual circuit is a path between points in a network that appears to be a discrete, physical path, but is actually a managed pool of circuit resources from which specific circuits are allocated as needed to meet traffic requirements. In this way, the software protocols of a network enable the two end points to communicate as though connected by a physical circuit. Network nodes provide the addressing information needed in the packets that carry the source data to the destination. Usually, a virtual circuit is cleared at the end of the call. However, it is possible for the virtual circuit to be left permanently established, so that a user who needs to communicate frequently with another user doesn't have to set up a virtual circuit for each call. This is known as a *Permanent Virtual Circuit* (PVC). The user is charged for this facility and for the quantity of data transferred.

The X.25 standard from the CCITT specifies a standard for the OSI network layer to be used by public data networks. It basically defines a connection oriented virtual circuit service similar to what has been described above. Being a layer three protocol, X.25 is directly comparable with the Internet Protocol of the Internet (which are described in Section 1.4).

In a typical implementation, intermediate nodes along a route will check for the occurrence of errors. This is usually done at the data link layer when the framing of the packet to be transmitted occurs and checksums are calculated. This carefully added redundancy enables nodes to detect errors and, depending on the algorithm used and the amount of redundancy, correct them. In case an error is detected and the integrity of a packet cannot safely be restored, the receiving node will typically ask the sending node to re-send the package. In this way, an end-to-end retransmission is avoided at the cost of longer processing times. If a network infrastructure can guarantee a very low incidence of error or data loss (this is, for instance, typical for a fiber optic medium), then error checking at intermediate points can be avoided. Error checking can then be performed by the higher layers end-to-end, and in the unlikely case that an error is detected, the packet will have to be retransmitted from the source node. This strategy will be faster assuming a very low probability of error, since processing times at intermediate nodes will be greatly reduced (checksum computation is typically a lengthy operation). Such techniques are known as *Fast Packet Transmission* (FPT). Due to the low incidence of error or data loss on currently used media, the FPT technique is gradually spreading. In this way, error checking is not provided at points along the route and the assurance that the packet has arrived without error is the responsibility of the receiver. Frame relay is a technology that uses principles similar to that of FPT. It provides for high performance digital transmission. Frame relay was defined in response to the increase in bursty data traffic, the need to interconnect Local Area Networks and the proliferation of client-server and distributed computing. As a telecommunication service, it is designed for cost-effective data transmission carrying intermittent traffic between LANs and between end points in a *Wide Area Network* (WAN). It puts data in a variable-size unit called a frame and leaves any necessary error correction (retransmission of data) up to the end points, which as explained, in the presence of high reliability links, speeds up overall data transmission. For most services, the network provides a PVC. An enterprise can select from different levels of service quality, prioritizing certain frames while marking others as less important. Packets are relayed at the data-link layer of the Open Systems Interconnection model. An important aspect of frame relay is that its frame can incorporate packets from different protocols such as Ethernet and X.25. The virtual circuits are permanently assigned. Frame relay relies on the customer equipment to perform end-to-end error correction. Each switch inside a frame relay network just relays the frame to the next switch. Current networks are sufficiently error free to move the burden of error correction to the end points. Frame relay complements and provides a mid-range service between ISDN and ATM, which operates in a somewhat similar fashion to frame relay. ATM is subsequently explained.

*Broadband networks* constitute an emerging concept changing the field of telecommunications. Research in this field was motivated by the growing market for multimedia computing applications. The term applies to telecommunication networks with bandwidth characteristics which allow them to transfer information at substantially higher rates than ISDN. Before ISDN had time to become widespread, technological developments in fiber optic and switching technologies opened the way towards broadband networks. The integration of digital transmission services in a broadband network is based on the B-ISDN concept. For B-ISDN the Asynchronous Transfer Mode has been agreed as the adopted method of realizing the broadband, integrated and, of course, digital aspects captured in the B-ISDN acronym. As a future network, ATM based B-ISDN will not only strive for common, ubiquitous access through the so-called Broadband User Network Interface, but also for a really integrated networking infrastructure. In other words, all users and services should exchange information through the use of ATM connections supporting ATM cells. The ATM is thus a key component of B-ISDN. The ATM is a dedicated, connection oriented, virtual circuit switching technology organizing digital data into fixed length packets (called cells) and transmitting them over a high reliability and throughput medium using digital signal technology. The processing of fixed length cells can be handled by hardware. A cell is processed asynchronously relative to other cells and is queued before being multiplexed over the line. ATM relies on different classes of service to accommodate different applications (voice, video, data). The increased throughput of ATM allows B-ISDN to offer a wider variety of services than do conventional communications networks. Transmissions over the B-ISDN can be conducted using either a *Constant Bit Rate* (CBR) or a *Variable Bit Rate* (VBR) service. CBR services are provided at a fixed transmission speed. VBR services can be offered for cases in which the amount of information to be transmitted is not constant. In videoconferencing, for example, the amount of information varies depending on the amount of movement within the pictures. ATM offers distinct, specific classes of quality of service for every application. The terminal through the use of signaling is typically able to indicate the requested quality of service class.

At the same time, the growing demand for mobility in services is urging towards wireless communications. Wireless services are extended to provide full terminal mobility with voice, data and video capabilities. Early deployment of *Mobile Services Networks* was based on narrowband networks. As multimedia applications extend to portable devices, broadband networks need to be better suited to support multimedia user requirements. The integration of broadband and wireless networks could provide nomadic access to the network and the infrastructure to support extended wireless services.

The *Global System for Mobile Communication* (GSM) is a digital circuit switching mobile telephone system that is widely used in Europe and other parts of the world. GSM uses a variation of *Time Division Multiple Access* (TDMA). It digitizes and compresses data, sending it down a channel with two other streams of user data, each in its own time slot. It operates at either the 900 MHz or the 1800 MHz frequency band.

The low data rate offered by GSM (9.6 Kbps) has led to the development of the *High Speed Circuit Switching Data* (HSCSD). HSCSD uses circuit switching wireless data transmission for mobile users at data rates four times higher than those of GSM. HSCSD offers a Quality of Service (QoS) comparable to that of many computer networks that use modems to communicate through the fixed telephone network. In this way, bandwidth demanding applications such as information services, large e-mail downloads, network games and live video camera surveillance can all be supported through a GSM network with a HSCSD capability.

GSM networks provide access to data services with a circuit switching data bearer service. Consequently, a connection will be in place for the total duration of a session, even when there is no data to send. Thus, scarce radio resources are wasted while even a busy user can take advantage of only one direction of the link. Moreover, billing depends exclusively upon the duration of the call and not on the amount of data. As data communications are typically bursty in nature, a circuit switching data connection could be established for each burst of data traffic. However, connection set up times are too long and signaling traffic is too high for this solution to be effective. It seems that there is no efficient way of handling frequent, small volume data transmissions, nor for handling bursty types of traffic within current GSM networks. Furthermore, connections are always dependant on circuit availability. Packet switching connection, on the other hand, using the *Internet Protocol* (IP), can provide a virtual connection to any other end point in the network. It can also provide alternative billing methods (pay-per-bit, pay-per-session, flat rate, asymmetric bandwidth).

The *General Packet Radio Services* (GPRS) enable cost-effective wireless access to bursty applications. It adds packet switching on top of the currently available circuit switched mode of operation. GPRS enables the introduction of packet data services, adding packet switching nodes to the existing GSM infrastructure. Thus, GPRS expedites mobile connectivity to the Internet or to a corporate Intranet.

The *Universal Mobile Telecommunications Service* (UMTS) is a third-generation broadband, packet based system used for the transmission of text, digitized voice, video, and multimedia at data rates up to and possibly higher than 11 Mbps, offering a consistent set of services to mobile computer and phone users. Based on the GSM communication standard, UMTS, endorsed by major standards bodies and manufacturers, is the planned standard for mobile users around the world by 2002. Once UMTS is fully implemented, computer and phone users can be constantly attached to the Internet while they move. Access is provided through a combination of terrestrial wireless and satellite transmissions. The higher bandwidth of UMTS will inevitably lead to the introduction of new mobile services.

## *1.3.2   Telecommunications services, signaling and management evolution*

The constant development of telecommunication networks and the services evolution are inextricably intertwined. Service demands initially provide the economic justification for investing in new, broadband networks. Once the network is in place, the need to exploit the investment appropriately pushes towards the introduction of new, advanced services. In effect, technical innovation is fueled by service demand and spawns new service demand in turn. This virtuous circle cannot possibly go on *ad infinitum*, but is likely to continue for many years to come, at least to the point where service demand is saturated. The point is that as a network is laid with the aim of providing certain services to users, it must also cater for each service's distinctive features. There are quality of service characteristics that are inherent in any type of service, and quality demands and user expectations vary accordingly. In a telephone conversation the transmission must be carried out in real-time, while in data communications the paramount consideration is the elimination of transmission errors. In the past, telecommunications networks have been designed to provide for one of these requirement sets. Today they can handle both.

The transition is as we noted, from single service networks to integrated services ones. Accordingly, the transport infrastructure must cater for the different requirements that are prevalent in each service. If the same network is used to transfer phone calls as well as medical images, is should be suitable for both reliable end-to-end transmission as well as real-time communication. This has become feasible due to both the enhancement of the available network resources and to the rapid telecommunications technology evolution. Fixed telephony, data transmission, multimedia and mobile communications are converging. Introducing intelligence in the network nodes with the form of more processing and more sophisticated protocols allows different routing decisions to be made, based on the type of service a given packet is used for. Although it did not fundamentally change the way services were provided, nor did it enable the provision of radically new services, the Intelligent Network concept represented a significant evolutionary step in that respect as it made it possible to introduce more intelligence in the network nodes without seriously affecting the switching systems. The intelligence required for the provision of a service was now placed in dedicated IN servers instead of every switch of the network. The switch functionality was restricted to basic call processing, and additionally, to the identification of IN service calls and to the routing of these calls to the IN servers (Venieris, 1998).

One of the first IN services to be introduced was the 800 service, which allows a called party reachable via an 800 number to pay for the cost of the incoming calls. It is used by large companies to offer a free information service to their customers. It has opened the way to a large number of services like reverse charging, call forwarding, or call deviation. IN services have the common characteristic of being accessible via a predefined numbering prefix. All public network operators have assigned each IN service a specific numbering prefix that enables the switching systems to distinguish between IN and normal calls and the intelligent service systems to distinguish among different services.

As services evolved, it became necessary for the user to communicate more complex information to the network. Accordingly, adjacent nodes in a network needed to exchange more information. Signaling is used for both these purposes. Signaling was initially in charge of the bearer connection control. In its early steps it had rudimentary capabilities, which were adequate for the services for which it was at that time used. Increasing service complexity has since then sustained a continuous signaling evolution. This evolution was fostered by advances in software engineering practices and methodologies. As signaling protocols become more complex, they are usually implemented along object oriented lines to facilitate software reuse and easier maintenance. Object orientation is examined in Chapter 3.

Signaling typically supports point-to-point, point-to-multipoint connections and mobility management operations. It also provides optimized connection handling procedures and guarantees efficient allocation of resources upon user request. Its role is substantial as it is in charge of the overall coordination of the network elements. Signaling is by definition relevant in networks providing connection oriented services with a form of circuit switching implementation, so it is of major significance in ATM or ISDN but almost non-existent in the Internet, as we explained in Section 1.2.

A distinction should be made between two different types of signaling. The first one is subscriber signaling, between terminal and network, at the *User Network Interface* (UNI). The other is interexchange signaling, between exchanges or switches, and pertains to the *Network to Node Interface* (NNI). The most important standardization bodies for ATM networks, the International Telecommunications Union (ITU) and the ATM Forum, issue recommendations for both UNI and NNI signaling standards.

The trend of networks to interconnect, yielding in this way bigger and potentially more unmanageable networks, underlined the need for effective network management procedures. Accordingly, network management has evolved. ISO has defined five key areas of network management: fault management, configuration management, security management, performance management, and accounting management. Management can formally be defined as the process of controlling a complex data network so as to maximize its efficiency and productivity. The more efficiently each of the above five areas of network management is implemented, the better the network is managed and administered overall. Since networks typically contain hundreds or thousands of elements that need to be monitored or actively managed, a tree structured *Management Information Base* (MIB) is used where management protocols can find all of the relevant information. At the top of this management tree lies the most general information available about a network. Each branch of the tree gets more detailed into a specific network area.

The two mainstream management protocols are the *Simple Network Management Protocol* (SNMP) and the *Common Management Information Protocol* (CMIP). Generally, SNMP works over the TCP/IP communication stack, while CMIP works over the OSI stack. SNMP came first and was commonly considered a Band-Aid solution to tackle management difficulties while other, more sophisticated, protocols were being designed. Indeed, two more advanced network management protocols emerged. The first was SNMPv2, which incorporated many of the features of the original SNMP as well as a few added features that addressed the original protocol's shortcomings. The second was CMIP, which was better organized and contained many more features than either SNMP v1 or v2. CMIP operates over the OSI network communication protocol stack. The difficulties in its implementation have hindered its widespread acceptance. CMIP was designed to build on SNMP by making up for SNMP's deficiencies and becoming a more refined and complete network management protocol. Its basic design is similar to that of SNMP, whereby *Protocol Data Units* (PDUs) are employed as variables to monitor a network. CMIP, however, contains 11 types of PDUs. The variables of CMIP not only relay information to and from the terminal, but they can also be used to perform tasks, i.e. they are more 'active' in a sense. CMIP has built-in security management devices that support authorization, access control, and security logs. On the other hand, CMIP's increased complexity resulted in a massive increase in necessary system resources. CMIP typically consumes more resources than SNMP by a factor of ten.

## 1.4 Internet Software Technologies

### *1.4.1 History of the Internet*

The Internet is a worldwide collection of interconnected government, education, and business computer networks – a veritable network of networks. Recently, the backbone of this network has been nearly totally commercialized, allowing it to be used for real, demanding business applications. The Internet originally took root as a research project, and much of its initial evolution was spurred by the academic community. Its commercialization is not unlikely to initiate a period marked by more aggressive marketing practices than in the past coupled with the placement of increased emphasis on Quality of Service aspects, necessary to capture lucrative market segments.

The Internet traces its origins in government financed research during the cold war era. For the increasingly sophisticated American military, developing a communication infrastructure capable of withstanding nuclear attacks that could wipe out some of its sites while continuing to be operational became important. In 1973, the *Advance Research Projects Administration* (ARPA) and the Department of Defense of the United States initiated a research program to investigate techniques and technologies for interlinking packet networks of various kinds. A key requirement was that if the most direct route was not available, routers would direct traffic around the network via alternate routes. Two networks or computers should maintain connectivity if at least one physical route existed (no matter how circuitous). The network designed could have no centralized structure since any hierarchical model would create a single point of failure. A redundant hierarchy would make it more difficult for an opponent to deal a devastating blow to the network, but it still fell short of providing the level of self-configurability and adaptability captured in the connectivity requirement. The Internet's decentralized, connectionless, flat structure stems directly from the extraordinary benchmark criteria it was designed to meet.

During the course of that research program, communication protocols had to be developed that would allow networked computers to communicate transparently across multiple packet networks. The project eventually became very successful, and there was increasing demand to use the network, so the American government separated military traffic from civilian research traffic, bridging the two by using common protocols to form an internet-work or *Internet*. The term 'Internet' is defined as 'a mechanism for connecting or bridging different networks so that two communities can mutually interconnect'. So, in the mid-1970s ARPA became interested in establishing a packet-switched network to provide communications between research institutions in the United States (ARPAnet). With the goal of heterogeneous connectivity in mind, ARPA funded research by Stanford University and Bolt, Beranek, and Newman to create an explicit series of communication protocols. The ARPA-developed technology included a set of network standards that specified the details of the protocols used by computers that would enable them to communicate, as well as a set of conventions for interconnecting networks and routing traffic. The result of this development effort, completed in the late 1970s, was the Internet suite of protocols.

Internet uses the TCP/IP protocol stack. TCP stands for Transmission Control Protocol and IP for Internet Protocol. We refer to any network using this stack as an internet, whereas the worldwide TCP/IP network is the Internet (capital 'I'). A person at a computer terminal or *Personal Computer* (PC), equipped with the proper software, communicates across the Internet by having the network driver place the data in an IP packet and addressing the packet to a particular destination on the Internet. The communication software in routers in the intervening networks between the source and destination networks reads the addresses on the packets moving through the Internet and forwards the packets towards their destination. The TCP protocol which is usually operating on top of IP guarantees end-to-end integrity, and allows it to differentiate between packets sent by two applications running on the same node.

Soon thereafter, there were a large number of computers and thousands of networks using TCP/IP, and it is from their interconnections that the modern Internet has emerged. While the ARPAnet was growing into a national network, researchers at Xerox Corporation's Palo Alto Research Center (PARC) were developing one of the technologies that would shape local area networking, namely Ethernet. Ethernet became one of the most important standards for implementing LANs. At about the same time, ARPA funded the integration of TCP/IP support into the version of the Unix operating system that the University of California at Berkeley was developing. As a result, when companies began marketing non-host-dependant workstations that ran Unix, TCP/IP was already built into the operating system software, and vendors such as Sun Microsystems included an Ethernet port on their devices. Consequently, TCP/IP over Ethernet became a common way for workstations to interconnect. The same technology that made PCs and workstations possible made it possible for vendors to offer relatively inexpensive add-on cards to allow a variety of PCs to connect to Ethernet LANs. Software vendors took the TCP/IP software from Berkeley Unix and ported it to the PC, making it possible for PCs and Unix machines to use the same protocol on the same network.

From a thousand or so networks in the mid-1980s, the Internet has grown to an estimated 1 million connected networks with about 100 million people having access to it (as of 1999). The majority of Internet users currently live in the United States and in Europe, but the Internet is expected to have ubiquitous global reach over the next few years. These numbers have continued to grow at geometric rates throughout the 1990s. There are now several thousand *Internet Service Providers* (ISPs), although the number is likely to decrease or maintain a modest growth over the next five years due to industry consolidation.

### 1.4.2    TCP/IP protocol suite

TCP/IP corresponds to the suite of protocols that form the basis of the Internet and intranets. These protocols define how data are subdivided into messages, and those in turn into packets for transmission across a network. These generic transport and routing facilities provide the bedrock upon which higher layer application protocols are founded. It is by using these higher layer protocols that applications can transfer files and send electronic mail. Most Internet users are familiar with services like e-mail (which uses the

*Simple Mail Transfer Protocol* – SMTP), World Wide Web (which uses the *Hyper Text Transfer Protocol* – HTTP) and FTP (which uses the *File Transfer Protocol* – FTP). SMTP, HTTP and FTP all have the TCP/IP protocol suite as a common denominator.

TCP and IP (Stevens, 1994) both rely on data encapsulation, data routing across networks, the exchange of information across non-uniform platforms, and fragmentation and reassembly of data. However, IP is a connectionless and unreliable protocol, while TCP is a connection-oriented and reliable one. While the TCP/IP protocol stack doesn't fit neatly into the seven layers of the OSI Reference Model (see Figure 1.4), it does provide all the necessary functionality for productive networking. Depending on where the distinguishing line between connectionless and connection-oriented protocols is drawn, TCP can also be considered a connectionless protocol (IP is in any case connectionless). That is, TCP treats each message sent to it by an application as a separate request and sets a new connection for it. This characteristic of TCP prevents the realization of an ATM-type connection in the Internet, since after the message is handled the connection is torn up. This allows bandwidth utilization by other hosts throughout the network, but makes it very difficult for the network to guarantee a certain QoS level. Under this, more demanding, interpretation of 'connection-orientation' TCP is still connection-oriented at the message level, since it treats all packets that have resulted from a message's breakdown as a set. Once the last packet of a message has been sent and acknowledged, the connection ceases to exist. The effect of TCP operating on top of IP is to wrap an unreliable channel and provide a reliable transport link on top of it.

The application protocols at the top of the protocol stack organize the data that networks will transfer. Beneath the application protocols is the transport layer, which is responsible for delivering the applications' information to the destination. Transport protocols in turn rely on the protocols found in the network (or inter-network) layer, which bear the responsibility of understanding the topology of the network and forwarding information through the network to its destination. The lowest layers deal with low level aspects of the network technology such as descriptions of the physical media and waveforms of electrical signals.

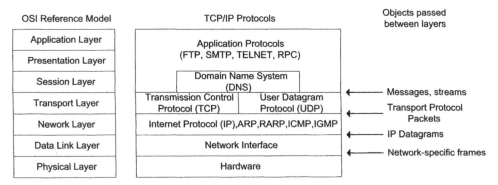

**Figure 1.4**  TCP/IP protocol stack

The protocols found in the network layer have the responsibility of forwarding packets across all the interfaces that make up an inter-network (which is one reason why IP is called the Internet Protocol) (Postel, 1981a). Thus, the network layer specifications include rules not only on how to forward packets, but also for describing the format of host addresses and defining protocols for delivering diagnostic information about the inter-network.

The Internet Protocol, is a connectionless protocol which defines the basic unit of data transfer used throughout a TCP/IP inter-network, choosing the path over which the data is sent. By convention, the packets that IP transfers are called *datagrams*. They consist of a datagram header and the data area. IP is used to transmit information from one machine to another. It is not able to discern between applications running on the same machine, however. Being connectionless, IP does not need to set up a connection with the machine before communicating with it. Being unreliable, it does not check the validity of the data it receives at the receiving end. IP performs two basic functions: packaging of data into IP datagrams that contain the destination IP address; and breaking up and reassembly of data that is too large to fit into a single IP datagram. The Internet Layer is only concerned with the destination of the packets it receives. Once it has resolved their location on the network, it ships the packets in the 'direction' of the appropriate host. This usually implies reading an IP datagram from one physical interface, processing it to resolve the destination address, and copying it on another interface which is determined by internally maintained lookup tables.

In addition to containing the network addresses of the source and destination of the datagram, the IP or datagram header includes options that can be used to define, for example, the type of service, route options, timestamp, and fragment processing options. Version 4 of IP (IPv4) has been used to provide the basic communication mechanisms of TCP/IP and the Internet over the last twenty years. However, as new applications for networks were introduced and inter-networks were greatly expanded (both in size and number), it became more obvious that a new version of IP was required. The proposed new version is called IPv6. Although, conceptually, IPv6 differs little from IPv4, many subtle details have changed. For example, IPv6 provides larger addresses and replaces the IPv4 variable-length options with a series of fixed-format headers (fixed format headers allow faster implementations for packet processing). The major changes in IPv6 are: addresses are now 128 bits long instead of 32 bits; a new flexible header format uses a fixed format header with a set of optional headers; network resources can be pre-allocated instead of using IPv4 type-of-service specification; IPv6 can more easily accommodate future extensions than IPv4.

Aside from moving data from source to destination, the protocols in the network layer co-ordinate functions such as discovering neighbors, controlling address assignments and managing group memberships. Neighbor discovery allows a system to identify other hosts and routers on its link. Hosts have to learn of at least one router so they can forward datagrams to systems not on their link.

In contrast to IP and related protocols in the network layer which are responsible for moving data from a source to its destination, the transport protocols (TCP and the *User Datagram Protocol* – UDP) have the responsibility of distinguishing among traffic from different applications within the system and compensating for IP's lack of reliability. Since unreliable links mean that some datagrams do not arrive at the destination, other methods are needed to provide reliability. In a process called *end-to-end verification* within the transport layer, the source and destination systems detect datagram losses and attempt to recover lost datagrams. The end-to-end software located within the transport layer uses checksums, acknowledgments, and timeouts to control the transmission of datagrams. Located above IP in the TCP/IP stack, either TCP (Postel, 1981b) or UDP (Postel, 1980) can be employed to determine the maximum transmission unit that can be used (that is, the packet size) and fine-tune transmissions accordingly.

The simplest level of service is datagram delivery, which does not guarantee delivery and does not protect against datagram corruption. UDP provides a best-effort delivery; it does its best to transfer datagrams for its applications, but makes no guarantees. Datagrams may get reordered or lost in transit, and the application using UDP must be prepared for such events. While the IP layer is responsible for transferring data between a pair of hosts (as identified by their IP addresses) on an inter-network, UDP must be able to differentiate among multiple sources or destinations within one host. Therefore the UDP protocol, in addition to the source and destination IP addresses, adds an additional field called the *port number* to differentiate among multiple applications on the same machine and to decide to which application to deliver a specific message. The main benefit of using UDP is its simplicity. Moreover, it is suitable for multicast transmissions since it is a connectionless protocol.

On the other hand, to achieve reliability, TCP uses a connection oriented service. TCP relies on IP address–port number pairs to identify individual connections. TCP's reliable delivery service offers error-free delivery, assured delivery, sequenced packets and no duplication of packets. As part of its operation, TCP divides a data stream into chunks called *TCP segments* and transmits them using IP. At the other end, TCP reassembles the TCP segments to form an uninterrupted data stream. TCP transmits data in segments encased in IP packets, along with checksums used to detect data corruption and sequence numbers to ensure an ordered byte stream. TCP is considered a reliable transport mechanism because it requires the receiving computer to acknowledge not only the receipt of data, but also its completeness, as well as provide for its correct sequencing. TCP is also able to distinguish among a number of applications using it on the same host, again by the inclusion of a port number. TCP offers a stream-like view of a connection, i.e. it does not preserve the boundaries of the messages it sends. If two messages are sent, an application reading the data on the destination host will not be able to recognize the number of messages that made up the byte stream it is receiving. This again contrasts with UDP, where message boundaries are preserved.

### 1.4.3    *Evolution of interprocess communication on the Internet*

Any consideration of computer communications software should take into account that the communicating applications can reside on either the same machine or on different machines across a network. The purpose of *Inter-Process Communication* (IPC) mechanisms is to define interfaces that will allow different processes to exchange data, and that will enable programmers to manage interactions between different processes. Among the most widely used IPC mechanisms are:

- Pipes
- Shared Memory
- Messages
- Sockets.

Some form of IPC is required to be provided at the operating system layer for all multiprocessing systems. For instance, messages between different processes are readily supported in the Windows operating system using the Win32 Application Programming Interface. Pipes are also supported in Unix systems as a one-way communication channel between different processes running on the same machine. Limiting our attention to networks, of more importance are IPC mechanisms which can be used between processes executing on different machines. Of the IPC mechanisms listed above, only sockets satisfy that requirement.

Sockets are basically an interface to transport layer protocols (TCP or UDP). TCP or UDP are protocols – sockets are simply an interface. This means that sockets may be used at the one end of a connection without necessarily being used at the other. While sockets are widely used on the Internet, the socket interface is suited to other networks as well. For instance, using the Socket 2 interface, one can write programs that run unmodified on top of the Internet as well as on top of ATM networks. Reflecting the association sockets have with the Internet, a socket can be configured to work with either streams or datagrams. In the first case TCP is used, while in the latter it is UDP. A program using sockets with UDP will have to concern itself with lost or out-of-order messages and take appropriate corrective action in any case. When using TCP, the programmer can rest on TCP's reliability mechanisms. TCP, however, tears down a connection once a message has been sent and reassembled at the receiving end. If a connection is used for a large number of short messages, then the expense of setting up a connection for stream communication may not be justified. Note that in this context, connection setup costs do not stem from committing resources along the route, but simply from setting up the necessary state information at the two endpoints (the execution of a handshake protocol usually precedes the actual transmission of data). In those cases, UDP communication may be more appropriate even in the face of the increased complexity the application will need to have.

While sockets can be used to create powerful networking applications, they place a significant burden on the programmer to define appropriate application-level protocols and to undergo nearly all data transformations himself. *Remote Procedure Call* (RPC) is a protocol that allows one program to request a service from a program located on another computer without having to understand network details. RPC uses the client/server model and allows communicating with a program on a remote process in much the same way that a call would be made in a local process. RPC's essential concept is hiding all the network code in the so-called 'stub' procedures that take care of all non-service-specific functionality. RPC does not add new capabilities over sockets, but it does make writing distributed applications easier (Birrell, 1984).

RPC, however, falls short of providing the facilities one has when making a local procedure call. For one thing, pass-by-reference is not permitted so the use of complex data types, such as user defined types, is difficult. The reason is that passing complex data types by copy (when pass-by-reference is not available) involves marshalling, and this presupposes a common, agreed-upon wire format. Specifying a uniform data representation standard to serve as the lacking wire format is a rather burdensome task as different machine architectures implement data in miscellaneous formats. Marshalling mechanisms, even when implemented, incur significant performance penalties as they are designed to cope with eventualities that may safely be dismissed in a low-level, specifically tailored, socket implementation. Furthermore, RPC couldn't compete in performance against more rudimentary mechanisms like plain sockets. A more severe handicap was the lack of security provisions, which meant that as RPC was executed across networks, the data carried was vulnerable to interception by potential eavesdroppers. Nevertheless, RPC as a networking middleware represented a major step forward, compared with other networking practices of the time, and paved the way for technologies like the *Common Object Request Broker Architecture* (CORBA) and *Distributed Component Object Model* (DCOM).

CORBA and DCOM are middleware technologies, not unlike RPC but significantly enhanced nonetheless. They are both designed to be language-independent, and while providing for procedural code as well, they seem particularly well suited for object oriented implementations. They have the twin goal of simplifying program development and maintenance and allowing objects installed on different computers to communicate as though they were located locally. Networking software development is targeted towards heterogeneous computing environments that integrate new systems with legacy components. The traditional two-tier client-server architecture is evolving towards the concept of a distributed systems architecture that enables invocations of operations on objects located anywhere on the network as if they were local to the applications using them. This distributed architecture can blur the distinction between 'client' and 'server' because the client components can also create objects that behave in server-like roles.

CORBA is a distributed object framework proposed by a consortium of industries called the Object Management Group (OMG). It specifies a system which provides interoperability between objects in a heterogeneous, distributed environment, and in a way transparent to the programmer. Common object semantics for specifying the externally visible characteristics of objects in a standard and implementation-independent way are defined. Clients request services from objects through a well-defined interface. A client accesses an object by issuing a request to that object. The request carries information about the service to be provided; this includes the name of the operation, the reference to the object that supports it, and the actual parameters. The object reference is an object name that defines an object reliably.

The main feature of the CORBA architecture is the *Object Request Broker* (ORB) that encompasses all of the communication infrastructure necessary to identify and locate objects, handle connection management and deliver data. A CORBA object is represented to the outside world as an interface with a set of methods. The client of a CORBA object acquires its object reference, and once it knows the type of the object and the method to be invoked, it can use the object reference as a handle to make the request, as if the object was located in its own address space. The ORB is responsible for all the mechanisms required to find the object's implementation, prepare it to receive the request, and to communicate the request to it. The object implementation interacts with the ORB through either the *Object Adapter* (OA) or through the ORB interface. CORBA is presented in more detail in Chapter 4.

DCOM is the distributed extension of the *Component Object Model* (COM). It basically adds an *Object Remote Procedure Call* (ORPC) layer on top of the *Distributed Computing Environment* (DCE) to support remote objects. COM components use standardized interfaces to pass data. As long as client or server objects implement the standard interfaces, the programming language used to create either of them is irrelevant. DCOM uses the same methodology to talk across networks that is used to perform interprocess communication on the same machine. A COM object can support multiple interfaces, each representing a different view or behavior of the object. An interface consists of a set of functionally related methods. A COM client interacts with a COM object by acquiring a pointer to one of the object's interfaces, and invoking methods through that pointer. Since the specification is at the binary level, it allows the integration of binary components possibly written in different programming languages.

Both DCOM and CORBA frameworks provide client-server types of communication. To request a service, a client invokes a method implemented by a remote object, which acts as the server in the client-server model. The service provided by the server is encapsulated in an object, and the interface of that object is described in an Interface Definition Language (IDL). The IDL interface serves as the contract between a server and its clients. Clients interact with a server by invoking methods defined in that IDL interface. The object implementation is hidden from the client. Some object oriented programming features are preserved at the IDL level, such as data encapsulation, polymorphism and single inheritance. CORBA supports multiple inheritance at the IDL level. In contrast, DCOM uses multiple interfaces to achieve the purpose of multiple inheritance (Box, 1997). DCOM, as well as a comparison between DCOM and CORBA, are presented in more detail in Chapter 4.

### 1.4.4   World Wide Web

On top of TCP/IP run a plethora of protocols. Associated applications employ these protocols to offer some tangible service to the end user. A multitude of services use the same lower layers, but each provides a different facet to the Internet. Users using a file transfer application perceive something entirely different than those using a mail client or a web browser. The *World Wide Web* (WWW) is perhaps the best known facet of the Internet. For most people, the Internet is the WWW. Technically, however, one could define WWW as the subset of Internet hosts that communicate using the HTTP protocol. HTTP is a typical client-server protocol, and so WWW can be considered as the world's largest client-server system. In fact, it's a bit worse than that, since all the servers and clients coexist on a single network at once. For the most part, there are many common paths to be shared between clients and servers, and it is not uncommon for two different clients to experience highly correlated quality of service characteristics. Client applications (Web browsers) are available that make accessing the WWW very easy. These client applications don't need to understand the complexity of the underlying network. From their point of view, it is a straightforward procedure to connect to and interact with one server at a time.

Initially, accessing the Web was a simple one-way process. Users made requests on a particular server and got as a response a file (called a Web page), which the browser software would interpret by rendering onto the screen of their local machine. The format of the file followed the syntax of a language specially developed for that purpose called the *Hyper Text Markup Language* (HTML). File pointers to images could be included in a Web page and later, pointers to sounds or video clips, making Web content more dynamic (and Web pages 'heavier'). However, people began wanting more than just accepting pages (no matter how fancy) from a server. They wanted full client-server capabilities so that the client could feed information back to the server, for example, to do database lookups on the server, to add new information to the server or to place an order (which required more security than the original systems offered).

The Web browsers and the invention of HTML were big steps forward: the main concept was that one piece of information could be displayed on any type of computer without change. The same HTML file (the same Web page) would be rendered differently to different machines depending on their presentation capabilities. However, browsers were still rather primitive and had to be rapidly improved as demands began to mount. They weren't particularly interactive and tended to overload both the server and the Internet, because any time a user needed to do something that required processing he had to send information back to the server for processing. It could take many seconds or minutes to find out that the user had misspelled something in his request. Since the browser was just a viewer it couldn't perform even the simplest computing tasks. One other problem was that as Web pages became heavier, downloading them consumed more bandwidth. A large percentage of the Internet's bandwidth was quickly found to be taken up by Web traffic, for instance by large image file downloads. Since IP cannot differentiate between Web traffic and, for instance, Telnet traffic, the entire Internet edifice was threatened, and not just the Web part of it.

To solve these problems, a number of approaches were taken. To begin with, graphics standards were enhanced to allow better animation and video within browsers. New formats were used that achieved far better compression rates. BMP images, which are notorious for their size, were replaced by JPG ones, which rely on decoding software on the browsers to render them. Since the scarcest resource was bandwidth and not CPU cycles on the terminal, it made sense to substitute one for the other. The computing problem (whereby clients forward all data for processing to the Web server) was more difficult to tackle. That problem could be solved only by incorporating the ability to run programs on the client end, inside the browser.

Although the Web's initial server/browser design allowed for interactive content, all interactivity was provided by the server. The server produced static pages for the client browser, which would simply interpret and display them. Basic HTML contains simple mechanisms for data gathering: text-entry boxes, check boxes, radio boxes, lists and drop-down lists, as well as a button that can only be programmed to reset the data on the form or submit the data on the form to the server. This submission passes through the *Common Gateway Interface* (CGI) provided on all Web servers. The text within the submission tells CGI what to do with it. The most common action is to run a program located on the server in a directory that's typically called 'cgi-bin'. Many powerful Web sites today are built entirely using the CGI approach, and end users can in fact do nearly anything with them. The problem is response time. The response of a CGI program depends upon how much data must be sent as well as the load on both the server and the Internet. The initial designers of the Web did not foresee how rapidly this bandwidth would be exhausted from the kind of applications people would develop. For example, any sort of dynamic graphing is nearly impossible to perform with consistency because a *Graphics Interchange Format* (GIF) file must be created and moved from the server to the client for each new version of the graph. The solution to this problem is to move some of the server's functionality into the client's PCs. So, instead of executing CGI programs in the server, the idea is to move this functionality to the client side.

It is useful here to note a theme that is underlying the entire Internet technology. Functionality exists only in the endpoints of the network, and not 'inside' the network (as it does, for instance, in PSTN). When the allocation of functionality between two end-points is out of balance, the remedy is simply to move functionality from one endpoint to another. It is not possible to 'install' functionality in the sub-network.

Returning to the CGI discussion, in light of the need to place more functionality in the browsers, four main technologies are relevant:

- *Scripting languages*: with a scripting language one embeds the source code for a client-side program directly into the HTML page, and the plug-in that interprets that language is automatically activated when the HTML page is being displayed. Scripting languages tend to be reasonably simple to understand, and because they have a simple textual form, they can be part of an HTML page and load very quickly as part of the single server hit required to produce that page. Scripting languages are really intended to solve specific types of problems, primarily the creation of richer and more interactive Graphical User Interfaces (GUIs). The most common scripting languages are JavaScript, VBScript (which looks like Visual Basic) and Tcl/Tk, which comes from the popular cross-platform GUI-building language.
- *Plug-ins*: one of the most significant steps forward in client-side programming was the development of the plug-in. In this way, a programmer can add new functionality to the browser by downloading a piece of code that plugs itself into the appropriate spot in the browser. It tells the browser 'from now on you can perform this new activity'. Very powerful behavior can be added to browsers via plug-ins (e.g. ability to display a video sequence), but writing a plug-in is a hard task. In addition to playing video clips, plug-ins can be put to a more subtle use: they can allow an experienced programmer to create his own scripting language and have it interpreted by the browser without needing to collaborate with the browser manufacturer. Thus, plug-ins provide a back door that allows the creation of new client-side scripting languages.
- *Java applets*: Java is a powerful programming language built to be secure, cross-platform and international. Java is being continuously extended to provide language features and libraries that elegantly handle problems which are difficult in traditional programming languages, such as multithreading, database access, network programming and distributed computing. Java allows client-side programming via the *applet*. An applet is a mini-program that will run only within a Web browser. The applet is downloaded automatically as part of a Web page (just as, for example, a graphic is automatically downloaded). When the applet is activated it executes a program. This is part of its beauty – it provides a way to automatically distribute the client software from the server at the time the user needs the client software, and no sooner. Users get the latest version of the client software every time they visit a page without difficult re-installation procedures. Because of the way Java is designed, the programmer needs to create only a single program, and that program automatically works with all computers that have browsers with built-in Java interpreters. One advantage a Java applet has over a scripted program is that it is in a compiled form, so the source code isn't available to the client. Since Java is a general purpose programming language, Java applets are generally more powerful then scripting language constructs can be. Java wasn't designed with the intention to be used in Web pages, so applet capabilities accrue very naturally from its overall architecture. Chapter 5 has more on the subject.
- *ActiveX controls*: Microsoft's ActiveX is the competitor to Java applets, although it takes a completely different approach. ActiveX was originally a Windows-only solution, but now is being developed by an independent consortium to become cross-platform. In the ActiveX approach, a program only needs to connect to its environment, and then can be incorporated into a Web page and run under a browser that supports ActiveX. Thus,

ActiveX does not constrain programmers to a particular language. Hence, an experienced Windows programmer can create ActiveX components using a programming language of choice such as C++, Visual Basic, or Borland's Delphi.

### 1.4.5    New trends

Packet switched networks, like the Internet, traditionally offer best-effort delivery of network traffic. However, this type of service is unsuitable for real-time applications such as interactive multimedia that often cannot tolerate retransmitted packets or indeterminate delays. If packet switched networks are to support real-time applications, a system other than best-effort delivery is required. In this sub-section, we briefly describe the methods that have been proposed for providing guaranteed service and QoS to interactive applications on IP networks.

Anticipating the variety of real-time applications and services that could be used on IP networks, the Internet Engineering Task Force (IETF) formed the *Integrated Services* (IntServ) Working Group (Braden, 1994;1997), which set out to design a set of extensions to the best-effort delivery model that is currently used on the Internet. This framework, the Integrated Services Architecture, provides special handling for certain types of traffic flows, and includes a mechanism for applications to choose between multiple levels of delivery services for their traffic. A fundamental principle of the Integrated Services Architecture is that network resources need to be controlled in order to deliver QoS, and that this requires the inclusion of admission controls. Along with this, there is a signaling protocol, the RSVP, that enables applications to send to the network requests with specific QoS parameters. RSVP operates on top of IP, occupying the place of a transport protocol in the protocol stack, but provides session layer services (that is, it does not transport any data). RSVP is an Internet control protocol like the *Internet Group Management Protocol* (IGMP) or *Internet Control Message Protocol* (ICMP), but it is not a routing protocol. It employs underlying routing protocols to determine where it should carry reservation requests. As routing paths change, RSVP adapts its reservation to new paths if reservations are not in place. The RSVP protocol is used by routers to deliver QoS control requests to all nodes along the paths of a flow, and to establish and maintain the state necessary to provide the requested service. After a reservation has been made, routers supporting RSVP determine the route and the QoS class for each incoming packet, and the scheduler makes forwarding decisions for every outgoing packet.

While many had hoped that RSVP would prove to be the mechanism for QoS provisioning over the Internet, a large contingent of Internet Service Providers have expressed concern over the amount of computational, memory, and other system resources which might be consumed by managing thousands or perhaps hundreds of thousands of flows. As a result, a small collective group within the IETF Integrated Services Working Group has been examining several methods to define less resource intensive mechanisms to provide differentiated services – somewhere between the traditional best effort model and the guaranteed and controlled-load service models defined in the Integrated Services architecture. Hence, a new architecture, known as *Differentiated Services* (DiffServ) (Blake, 1998), (Nichols, 1998;1997) was created which provides reasonable levels of differentiation, but which does not yet impose the path and resource reservation state maintenance required by RSVP. Without adding a new signaling protocol (which has, in and of itself, been a topic of much dissension), the most logical place to consider placing indicators for differentiation is the IP header. If we examine the IP header, we find the 8 bit 'Type Of Service' field which could indeed be used to convey differential service information. This is due to the fact that none of the bits within this particular field have been widely used for anything else to-date. The Differentiated Services model defines a set of traffic classes, each of which is designed to serve applications with similar QoS demands. A traffic class describes the *Per Hop Behavior* (PHB) that the packets of this class should experience in each network node. The per hop behavior determines the priority, the maximum delay in the transmission queues, the link-sharing bandwidth and the probability of a packet to be dropped. The DiffServ model ensures high scalability by separating the operations performed in the borders of the network from those accomplished in the core network. Border routers perform more complex tasks such as traffic policing and shaping, marking and prioritized routing, as Figure 1.5(b) shows. Marking is the process of classifying IP packets belonging to a specific IP flow and assigning them to the appropriate traffic class. All of the above operations are performed on a per flow basis as in the IntServ model. However, the small number of active IP flows handled by a border router does not cause the scalability problems that exist in the IntServ architecture. On the other hand, core routers carry out only one simple task, that is, prioritized routing.

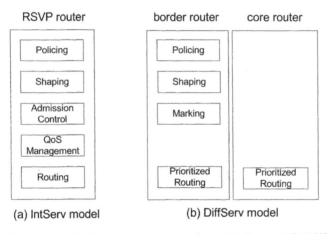

(a) IntServ model          (b) DiffServ model

**Figure 1.5**  Tasks performed in the network elements of an (a) IntServ and (b) DiffServ network

DiffServ core routers do not keep any information for the established IP flows. On the contrary, they simply serve packets according to the traffic class that the Integrated Services border router has assigned to them. Hence, each DiffServ core router has to know only the number of traffic classes and the corresponding per hop behavior of each class.

Besides the Integrated and Differentiated Services, there are also other approaches for QoS provisioning over the Internet, the most noticeable being the Layer 3 switching. Notable vendor specific solutions include Cisco's NetFlow and Tag switching technology (Rekhter, 1997) and Ipsilon's IP switching technology (Newman, 1997). Standards-based solutions include *Multi-Protocol Over ATM* (MPOA) (ATM Forum, 1997) and the *Multi-Protocol Label Switching* (MPLS) (Rosen, 1999), which is the evolving specification for cohesive Layer 2 and Layer 3 switching and routing. The target of all of the above technologies is to utilize the characteristics of the physical layer, which in many cases can be ATM, and to move operations from layer 3 (network layer) to layer 2 (link layer), leading to increased performance. As an example, we briefly describe the basic operation of MPLS. As a packet of a connectionless network layer protocol travels from one router to the next, each router makes an independent forwarding decision for that packet. That is, each router analyses the packet's header, and each router runs a network layer routing algorithm. Each router independently chooses a next hop for the packet, based on its analysis of the packet's header, and the results obtained from running the routing algorithm. Packet headers contain considerably more information than is needed to simply choose the next hop. Choosing the next hop can therefore be thought of as the composition of two functions. The first function partitions the entire set of possible packets into a set of Forwarding Equivalence Classes (FECs). The second maps each FEC to a next hop. As far as the forwarding decision is concerned, different packets which get mapped into the same FEC are indistinguishable. All packets which belong to a particular FEC and which travel from a particular node will follow the same path (or if certain kinds of multi-path routing are in use, they will all follow one of a set of paths associated with the FEC). In conventional IP forwarding, a particular router will typically consider two packets to be in the same FEC if there is some address prefix X in that router's routing tables such that X is the 'longest match' for each packet's destination address. As the packet traverses the network, each hop in turn re-examines the packet and assigns it to a FEC. In MPLS, the assignment of a particular packet to a particular FEC is done just once, as the packet enters the network. The FEC to which the packet is assigned is encoded as a short fixed length value known as a 'label'. When a packet is forwarded to its next hop, the label is sent along with it, that is, the packets are 'labeled' before they are forwarded. At subsequent hops there is no further analysis of the packet's network layer header. Rather, the label is used as an index into a table which specifies the next hop, and a new label. The old label is replaced with the new label, and the packet is forwarded to its next hop.

# References

ATM Forum (1997) Multiprotocol over ATM Specification, version 1.0. ATM Forum.

Birrell, A. and Nelson, B. J. (1984) Implementing Remote Procedure Calls. *ACM Transactions on Computer Systems*, **2**(1).

Blake, S. et al. (1998) An Architecture for Differentiated Services RFC 2475. IETF.

Box, D. (1997) Q&A ActiveX/COM. *Microsoft Systems Journal*.

Braden, R. et al. (1997) Resource Reservation Protocol (RSVP) – Version 1 Functional Specification. IETF.

Braden, R., Clark, D. and Shenker, S. (1994) Integrated services in the Internet architecture: an overview RFC 1633. IETF.

Chapman, M. and Montesi, S. (1995) Overall Concepts and Principles of TINA. TINA-C deliverable.

Halsall, F. (1995) *Data Communications, Computer Networks and Open Systems*. Addison-Wesley.

Newman, P., Minshall, G. and Lyon, T. (1997) IP Switching: ATM Under IP. *IEEE/ACM Transactions on Networking*, **6**(2).

Nichols, K. et al. (1997) A Two-bit Differentiated Services Architecture for the Internet. Internet Draft <draft-nichols-diff-svc-arch-00.txt> available from: ftp://ftp.ee.lbl.gov/papers/dsarch.pdf.

Nichols, K. et al. (1998) Definition of the Differentiated Services Field (DS Field) in the IPv4 and

IPv6 Headers RFC 2474. IETF.

OMG (1997) Interworking between CORBA and IN Systems. Request For Proposals, OMG.

Pan, H. (1998) *SNMP-based ATM Network Management*. Artech House.

Postel, J.B. (1980) User Datagram Protocol RFC 768. IETF.

Postel, J.B. ed. (1981a) Internet Protocol RFC 791. IETF.

Postel, J.B. ed. (1981b) Transmission Control Protocol RFC 793. IETF.

Rekhter, Y. et al. (1997) Cisco System's Tag Switching Overview RFC 2105. IETF.

Rosen, E., Viswanathan, A. and Callon R. (1999) Multi-Protocol Label Switching Architecture, internet draft. IETF.

Stallings, W. (1993) *SNMP, SNMPv2, and CMIP*. Addison-Wesley.

Stevens, W.R. (1990) *Unix Network Programming*. Prentice Hall.

Stevens, W.R. (1994) *TCP/IP Illustrated Volume 1 –The protocols*. Addison-Wesley.

Tanenbaum, A.S. (1989) *Computer Networks*. Prentice Hall.

Venieris, I.S. and Hussmann, H. eds. (1998) *Intelligent Broadband Networks*. John Wiley & Sons Ltd.

# 2

# Future Trends in Telecommunications Software Technologies

## 2.1  Software in Telecommunication Environments

Telecommunication networks are fundamentally complex distributed systems. In fact, *Public Switched Telephone Networks* (PSTN), with their hierarchies of local exchanges, tandem offices and higher level regional or national offices, can be regarded as the first distributed computing systems. Their distribution was structured, however, and varied depending on the hierarchical level examined. Moreover, irrespective of the distribution of control, they had an inherently centralized structure in terms of the management operations and of the service model employed. Meszaros (1999) notes that distribution was propelled by the need for increased reliability and high capacity, and that from a certain point onward, successfully managing and handling this complexity became essential in order to ensure that overall reliability was increased rather then decreased.

Telecommunication software has in many respects certain particular characteristics of its own. First of all, it is usually embedded, implying strong interaction with the hardware it controls. In the other direction, the hardware usually imposes limits on what the software is allowed to do. Secondly, telecommunication software is used in most cases for real-time, stimulus-response systems. That is, the systems are reactive: they receive a message and proceed to give a response or take some action according to the message's contents. Thirdly, telecommunication software is usually parallel and distributed. There is seldom a single 'main' program. More often there is a mesh of processes, usually implementing certain finite state machines, that communicate by message-passing according to specified protocols. Finally, implementation requirements for telecommunication software are usually very precise and unambiguous, in contrast to other types of software where it is not unusual for requirements to change during implementation. These characteristics, pertinent to telecommunication software, have in turn shaped the telecommunication networks themselves. The availability of precise specifications and the *a priori*, detailed knowledge

---

*Object Oriented Software Technologies in Telecommunications,* Edited by I. Venieris, F. Zizza and T. Magedanz
© 2000 John Wiley & Sons Ltd

of the software's task have encouraged the placement of much functionality into the network itself, which was much more capable of supporting the required processing than were impoverished terminals at the network's edges. Telecommunication networks are accordingly very 'active', and the level of involvement they assume, to a large extend, explains their inflexibility.

Computer networks, on the other hand, view the 'subnetwork' as a passive channel used to convey data between intelligent nodes that are located at the network's perimeter. Data are opaque to the intermediate nodes that convey them, and are interpreted only at the end nodes. Telecommunication networks assume terminal equipment of very modest processing capabilities, and view the network as playing an active role in providing, managing, and maintaining services. Indeed, the concept that the network offers a service contrasts with the computer networks notion of services being offered at the endpoints of the network.

## 2.2    The Role of Services in Telecommunications

Services are an intrinsic part of a communication network. The entire network architecture revolves around them, seeks to facilitate their speedier development, ease their introduction or withdrawal, support their billing and otherwise supervise their operation. The set of services to be offered has a profound impact on the design of the network architecture. Services are not viewed simply as value adding applications to be composed using more elementary facilities available underneath and to operate on top of a general-purpose communications subsystem. That conception would have favored a generic network infrastructure with a large set of simple facilities which would be employed by actual services; the more complex a service is, the greater the set of elementary facilities it would use and the more intricate the interactions between them it would prescribe; nevertheless, the emphasis would be placed on finding a set of elementary facilities that would lend themselves to systematic arrangements and would form a coherent framework for the composition of more complex services. Real telecommunication networks, however, are seldom generic; they are developed with specific services in mind and their structuring, protocols, and the procedures and concepts they define reflect this. It is in this respect that traditional telecommunication networks are often described as monolithic. They resemble a very large, finely-tuned mechanism aimed at providing a small set of services. In designing and implementing this mechanism, performance gain opportunities through tighter integration were pursued. Maintaining a layered architecture, such as that formulated by the OSI reference model described in Chapter 1, with very specific interfaces through which information between different layers would be communicated, was an objective – albeit one that was dispensable when in conflict with more concrete and measurable ones. The reader of telecommunication specifications is often surprised to see how 'low' into a network's protocol stack specific service-related concepts are still identifiable. The final aim of the architecture (the services to be provided) thus becomes the predominant concern at every level of it. Everything is geared towards providing the actual services. Protocols are defined so as to be useful instruments to the set of services that are expected to use them. Often, protocols provide composite operations that would be best implemented in higher layers in a perhaps less succinct but more modular manner. Essentially, current and past practices correspond to doing away with layering and having high-level, service-related concerns suffused over the entire architecture. The resultant network will certainly be efficient, concerning the services in mind. It will probably have performance advantages over less vertically integrated architectures. However, its ability to incorporate new services, especially those corresponding to different paradigms of provisioning or of interaction, will be fundamentally marred.

These properties of telecommunication networks are, to a large extent, remnants of their electromechanical past. To offer a *Plain Old Telephony Service* (POTS), a telecommunication network had to allocate and reserve resources across voice trunks passing through a number of switches. Protocols were defined between adjacent switches and at the interface between the home appliance and the local exchange to facilitate communicating the necessary information. First implementations were entirely hardware based, where electric pulses steered electromechanical switches to select the appropriate outgoing circuit. It was not until these systems were replaced by programmable computers and electrical pulses by 'proper' signaling protocols that it became feasible to design a network in a more flexible and adaptable manner. The ability to introduce new services was largely dependant on very mundane aspects of the transport network's technology. For instance, the 800 service (the Toll-free or Freephone service) couldn't have been implemented with the old system of using pulses to convey signaling information. When multifrequency tones, and later, digital messaging, were employed for the same purpose, it became possible to transmit more information that allowed the 800 service to be implemented. For instance, signaling the whereabouts of the calling party relative to the ultimate destination provided enough information to the terminating exchange to generate the appropriate billing (Meszaros, 1999). Throughout, a strong intertwining is shown between services offered and low-level aspects of the network's operation. As we have explained, that this correlation is still observed is indicative of the effect old paradigms are having on current telecommunication systems that are no longer technically bound to adhere to the constraints of the past.

## 2.3    The Role of Services in Computer Networks

In contrast, services are far more volatile entities with respect to computer networks. There is no such thing as introducing a service 'to the network'. The network is pretty much generic, being more or less a dummy conveyor of bytes. Service-related intelligence and processing lie outside the network, not inside it. The operation of the services is completely transparent to the network itself. Accordingly, the impact that services have on the design and implementation of computer network architectures is significantly less than in telecom architectures. The fact therefore remains that computer networks are much less 'intelligent' than telecom ones, and at the same time much more generic.

These characteristics of computer networks, as contrasted with those of telecommunication ones, can help explain the different kinds of problems each faces, and the different merits for which each is lauded.

## 2.4    Relative Pros and Cons of the Telecommunication Networks Approach

Telecommunication networks, having being designed and implemented with specific services in mind, are more adept at handling service-specific issues such as authentication or billing. They can provide *Quality of Service* (QoS) guarantees more easily, since the protocols they employ can understand and communicate such considerations. QoS can more effectively be provided at the lower layers of an architecture (near the network layer). Therefore, the fact that lower levels were built with specific service paradigms in mind (and with an understanding of the needs they will pose and the facilities they will expect from lower layers) makes it easier to express service requirements in terms of lower protocol parameters. That is, the necessary translation between user-perceivable quality characteristics of a service and the transport protocol interactions or parameters that provide for those comes more naturally when there are synergies to be exploited between lower and higher level protocols and services. In a monolithic and tightly integrated system, such synergies are pursued, as we have noted, to the detriment perhaps of its layering structure. Nevertheless, they yield a more efficient architecture for the intended services. It should therefore come as no surprise that the price to be paid for this increased efficiency is cumbersome introduction of new services. Particularly in cases where a new service is characterized by a paradigm essentially different than that for which the architecture caters, the introduction of the new service can be very problematic. This trade-off between efficiency and flexibility is a recurring theme in many of the chapters of this book. For instance, the telephone network offers a mere switching facility and is not suited for introducing new routing, charging, and number translation capabilities in a stand-alone fashion. While the availability of program controlled switches at the end of the 1960s opened the way to the introduction of switch-based services, at the same time it accentuated the limits of service introduction in these networks. Supplementary service provision was shown to require the modification and enrichment of the basic call process in all the nodes supporting the new functionality. The high cost of the provision of new supplementary services to a large number of subscribers in a multi-vendor environment (a large number of switches needed to be modified) demonstrated the inefficiencies of service introduction conclusively. The Intelligent Network idea (see Chapter 7) has sought, with considerable success, to ameliorate the situation, but is in turn subject to its own vices and does not change the service provisioning paradigm dramatically. Nevertheless, there is significant room in it for improvement, and as it was the most promising technology to enter the mainstream of telecommunication practices for quite some time, it has been the focus of much research effort.

## 2.5    Historical Practices that Underpinned Differentiation

Computer networks are on the antipodes. The Internet had the fortune to evolve with the aim of providing a data transfer service which is as generic a service as one can be. Being the product of academic research it lacked a centralized business model, and was much more distributed in nature than telecom networks. The lack of a centralized management and operations model, and the placement of most necessary intelligence at the end points of the network, gave computer networks the edge in introducing new services quickly. It is much faster to change a single node and install a new version of a service there than to upgrade an entire array of switched or intermediate systems that collaborate to provide a service and between which multiple dependencies may hold. Data communications were, from the beginning, implemented in software (usually only the physical and data link layers were implemented in hardware). A software implementation makes it easier to come up with a layered architecture than an electromechanical or hardwired one does. In fact, in the latter case it is more likely that a layering architecture is altogether infeasible. It is also possible that even implicitly, software engineers were more inclined towards producing an implementation consistent with the goals of functional encapsulation, abstraction of details and observance of the conceptual model than did telecommunication engineers. This is not unreasonable to argue given the different background each group had. Also, committees specifying telecommunication protocols in most cases had a very good idea of the kind of applications that will employ these. It was thus easier to attempt to employ this knowledge for the purpose of creating efficiency, even if that was not opportune for the purposes of modularity and flexibility. This was not always the case in protocol specification for computer networks. Computer networks have always been in a state of constant evolution, and were not characterized by entrenched notions on how they will be managed, operated, and what kind of services they would offer. It is conceivable that this uncertainty emphasized strict layering as the only resource available to guard against the adverse effects of this fluidity. Layering was the only way to guarantee that a certain implementation will retain its applicability even in the face of a continuously changing environment. This in turn encouraged a more cautious approach that assumed very little. There were fewer incentives to vertically integrate protocol stacks and produce more monolithic architectures for exactly the same reason.

## 2.6    An Academic Perspective

One very important paper that explicitly referred to issues of this kind was by Saltzer (1984). In it, the authors defined a design principle (called the 'end-to-end argument'), with the intention of providing a set of guidelines when deciding on the placement of functions among the modules of a distributed computer system. Their central proposition (embodied in the design principle they proposed) was that functions placed at low levels of a system may be redundant or of little value when compared against the cost of providing them at that low level. In that article, the authors examined several individual cases, from bit error recovery and security using encryption, to delivery acknowledgement and duplicate message suppression. We will briefly reproduce here their discussion on the 'careful file transfer' problem to help readers gain an understanding of the kind of considerations and debate that marked that period, and compare this with equivalent problems that faced telecommunication networks and the sort of approaches that finally dominated there. Although the example offered here may seem too narrowly defined and lacking in scope, the ideas that are developed have the widest range of application, as will be shown.

The careful file transfer problem describes the plight of an imaginary engineer responsible for designing a file transfer application to 'carefully' transfer a file between two computers A and B linked with a data communications network. 'Careful' alludes to the requirement that the engineer should not content himself simply with a 'best effort' service, but must guard against failures that can occur at various points along the way. It is assumed that the engineer takes nothing for granted and is responsible for defining and implementing lower layer protocols in any way he sees fit.

From the file transfer program's point of view, it has to: read the contents of the file from the file system in fixed, disk-format independent blocks; pass it to the communication subsystem in fixed or variable size packets, generally different from file block sizes; rely on the communication subsystem to correctly move packets between computers A and B; and rely on the peer file transfer application at computer B to correctly reverse the procedure and write the file to disk. Threats to this transaction are:

- The file representation at A's hard disk may be corrupted to start with.
- The software of the file system, the file transfer program, or the data communication system might make a mistake in buffering and copying the data of the file, either at host A or host B.
- Transient errors can occur at either host.
- Individual bits along their way may be changed or packets may be discarded, delivered out of order or duplicated.

Two approaches towards dealing with this problem are conceivable: one is to attempt to reinforce each step along the way at the pertinent layer. This may include time-out mechanisms at the communications subsystem, checksums stored with each file to thwart bit errors with redundancy, or even such brute force countermeasures as doing everything two or three times. The alternate approach, called by the authors 'end-to-end check and retry', depends simply upon the end applications to make all necessary checks and simply repeat the entire transaction if an error is detected. End-to-end check and retry is thus contrasted with, for instance, router to router check and retry. In the latter case, checks, retries and timeouts are associated with individual packets between each hop in an attempt to increase the reliability of the communication subsystem itself.

The end-to-end argument posits that, in general, many functions, contemplated for implementation in lower layers of the system, are best implemented only with the knowledge and help of an application standing at the end points of the communication system. Therefore, the argument goes, providing that questioned function as a feature of the communication system itself:

- is costly, especially if unreasonable high levels of reliability are pursued; and it
- does not absolve the end application from performing most of these checks anyway, since failures can also occur outside the scope of the communications subsystem.

In short, the argument's reasoning is against low level function implementation, as it advises the network designer to resist the temptation to 'help' users and end applications by taking on more functions that necessary. The authors however conceded to the fact that a low level implementation may be appropriate on performance grounds and advised that decisions of this kind merit close inspection.

The end-to-end argument is a simple but profound idea. As such it is applicable to many disciplines besides the one for which it was originally articulated. In the original paper, applications of the argument were presented at the design of *Reduced Instruction Set Computers* (RISC), and even at the high-level auditing procedures of the banking system. For the purposes of our discussion, it provides great insight into the subtle differences in approaches, conceptualization and aims that resulted in such noticeable differences between data communications and telecommunications. It can also serve to identify opportunities for convergence at various levels and anticipate paradigm shifts required to effect them.

## 2.7    Computer Networks Revised

Internet and computer networks in general have, for the most part, adhered to the end-to-end argument's tenet in contrast to telecommunication ones. It can be argued, however, that even for computer networks, a point is reached where the performance gains to be exploited by making the network marginally more intelligent (i.e. service aware) would warrant placing more functionality inside the network. In fact, as terminal capabilities evolved both in terms of their computing power and their presentation capabilities, new services became conceivable. Most of them had a definite multimedia character and included, for instance, audio and/or video transmission while involving in certain cases more than one party. Bandwidth consumption reached unprecedented heights, and along with the explosive growth of the Internet, placed unanticipated demands on the infrastructure available. The question of how to use the available bandwidth more effectively nudged its way to the foreground because of the severity of the problem. As it turns out, efficient use of available resources sits oddly with a service-unaware network. An example might serve to illustrate this.

A very popular format for video streams is *Moving Picture Experts Group* (MPEG). MPEG exploits spatial and temporal redundancies in a typical video sequence to produce a more condensed, encoded representation of a video stream's frames. Spatial similarities, for instance, represent a large percentage of a typical video frame having large areas of roughly the same color (e.g. the sky). Temporal similarities correspond to each frame of a video having areas similar to the previous (e.g. background that remains static). By exploiting the fact that we know in advance that both these kinds of similarities abound in typical video streams, a compression algorithm can be defined by taking these into account. MPEG does exactly that. The basic scheme is to predict motion from frame to frame in the temporal direction (inter-frame compression), and then to use discrete cosine transforms to organize the redundancy in the spatial directions (intra-frame compression). The MPEG standard describes three different types of frames used for the prediction of subsequent frames. The point is that certain types of frames are more important than others in reconstructing the video stream. Therefore, end-to-end reception would be much improved if, in the case of network congestion, intermediate routers were dropping 'B' or 'P' frames instead of 'I' ones. This is a very indicative example of the sort of service awareness built inside the network that could be profitably employed by end applications. It is also an often cited example when advocating active network concepts. The proposal of a *Resource Reservation Protocol* (RSVP) or the Differentiated Services approach, to incorporate quality of service mechanisms in the Internet protocol stack, is not inherently different to proposing that a router be able to distinguish packets carrying 'I' frames from those carrying 'B' ones. It is in fact a step in the same direction. The differentiated services approach requests that Internet protocols are augmented in such a manner as to allow end applications to communicate to intermediate routers a partitioning of the traffic they generate into different classes. Routers can then be discriminatory in their discarding operations favoring certain classes of packets over others. RSVP, which was described in more detail in Chapter 1, can be very simply thought of as a more fine-grained (and complex) variant of differentiated services (although it predated the latter).

In the light of our previous discussion, such efforts correspond to aiming for an Internet which would be more service aware by implementing common functions inside the 'communications subsystem' instead of the end applications. That the need to make a compromise, when there are significant performance opportunities, was in fact acknowledged in the original 'end-to-end argument' paper is a further attestation to its continued relevance.

The trend is two-fold: viewing datacoms and telecoms at the opposing sides of a continuous spectrum naturally raises the question of whether any of these technologies could be benefited by moving towards a more balanced position where the associated properties could yield more benefits. From their part, telecommunication networks can also change by trying to become more generic: this would incur some performance penalties, but the gains accruing from the expeditious deployment of new services and a more flexible and evolvable architecture might offset those. Computer networks, on the other hand, might incorporate more service-specific functionality into their core (RSVP is only one example). This could compromise the layering integrity, since it could lead to functionality being placed lower than a rigid layering would warrant. It could also lead to redundancy, since some of these functions may in any case need to be re-implemented at higher layers. On the other hand, moving functionality into the core will certainly buy some efficiency and performance bonuses. Making computer networks less generic than they are today will probably make service introduction costs marginally higher, but it will also allow the network to be more aware of the services that are using it, thus allowing it to correlate and utilize its resources more efficiently. The MPEG example is very demonstrative of the effect that a more intelligent utilization of the network resources can have for the end applications. However, moves can also be identified that are not always in the 'convergence direction', as it were (see Figure 2.1).

For instance, proponents of the Active Network idea, who advocate applying active network principles in computer networks and in the Internet, clearly see some room for improvement as regards the degree of flexibility demonstrated by the present day Internet, and wish to make its basic infrastructure, its routers and the respective protocols even more generic. While one of the aims of Active Networks is a more efficient provision of services – through a more 'service aware' allocation of resources – the intention is not to encompass service-specific functionality in the transport network. It is rather to remove all but the most rudimentary functionality out of the 'subnetwork' while providing for hooks that would allow *ad hoc* facilities to be injected and offered to the services above, in a dynamic manner. The infrastructure will become less service aware, not more so, as it will be stripped of any functionality that cannot be safely 'engraved' in the infrastructure. Of course, the infrastructure that will be composed at runtime above the previous one (by means of capsules, etc.) will be more service aware. So, to an extent, deciding whether Active Networks represent a more or less service aware infrastructure actually depends upon what the definition of 'infrastructure' is. Active Networks in this sense appear to satisfy the apparently conflicting goals of a more service aware and, at the same time, more generic infrastructure. The Active Networks idea is presented in more detail in Chapter 6.

In contrast to the Active Networks approach, standardization efforts aiming at introducing QoS characteristics on the Internet can be viewed as using different, more traditional means to pursue the same goal of more efficient service provision. Here, instead of making the network and transport layers even more generic, it is actually attempted to make them the opposite, that is, to place a complex (in the case of RSVP) algorithm right into the routers providing such QoS facilities that the actual services could count and leverage on. Not inappropriately, RSVP has been described as the Internet's signaling, which at least helps to visualize its role using common metaphors. The fact that the Internet has a routing, connectionless transport network whereas telecommunication networks usually have switched ones is not entirely unrelated to the fact that the Internet's structure is much more 'flat' and decentralized. Switched systems usually had preconfigured, fixed-bandwidth circuits, and the bandwidth allocation granularity is commonly very coarse. Even when signaling protocols are employed, a virtual circuit is set end-to-end and individual switches are instructed to steer the bitstreams accordingly. This in general entails high maintenance costs and common protocols to which switching equipment would adhere. Routing, in contrast to switching, is not required to set up an end-to-end circuit, and individual routers do not collaborate as much as do switches participating in a virtual circuit. This fits well with the Internet's decentralized model, but in turn exacerbates the problem of successfully co-ordinating network resources. Internet signaling in the sense of RSVP attempts to improve the situation. One way or another, the addition of QoS functionality can do much to help multimedia services operate with the kind of efficiency and reliability that characterizes telecommunication networks. Not surprisingly, the more facilities that are provided by the network, the harder it will be to cater for services having radically different requirements as to the type of facilities they expect. The addition of QoS into the Internet will move it closer towards telecommunication networks as they are today, while the prevalence of ideas supported by the Active Networks community will move it further away.

## 2.8    Telecommunications Revised

The situation is nearly symmetrical in the case of telecommunication networks. Nearly, since we are not aware of any intentions to make telecommunication networks more monolithic or tightly integrated in order to pursue any, as yet, elusive performance gains. On the contrary, the predominant concerns are that telecommunication architectures are overly monolithic, rigid and being tightly integrated, abound with dependencies that impede the introduction of new services and impose daunting management and control tasks on the network administration. Those who share these views argue that telecommunication networks have only to benefit by embracing some of the characteristics that have contributed to the unparalleled growth of the Internet. More specifically, that flexibility should be pursued even if it is not advantageous from a performance point of view; that service-transport network dependencies should be eradicated; that a rigidly layered architecture should be defined and implemented; and that relative protocols and the functionality they offer should be confined to the set of considerations that can be shown to be appropriate at each layer. The *Telecommunications Information Network Architecture Consortium* (TINA-C) has perhaps gathered the greatest momentum in advancing these goals (see Chapter 4). The initiative was established with the intention of defining and validating an open architecture for telecommunication services for the emerging broadband networks. TINA aims at delivering a consistent architecture for complex, distributed network management components and applications. From an architectural point of view, it strives towards resolving heterogeneity and providing a single, flat communication space. A flat communication infrastructure should be contrasted with the massively redundant, hierarchical ones which are now prevalent in telecommunication environments, to obtain a feeling as to the direction TINA is pursuing. Not surprisingly, the TINA consortium explicitly describes its architecture as one based on the following principles: object oriented analysis and design, distribution, decoupling of software components, separation of concern. TINA's goals are quite momentous and will amount to a total re-engineering of current networks. Much effort has been directed towards finding optimum evolution or migration paths that will allow a step-wise approach while capitalizing as much as possible (and for as long as possible) on already deployed equipment and software. For this reason, TINA is embracing the *Common Object Request Broker Architecture* (CORBA) for the implementation of its single communication space. CORBA is widely used in computer networks, but applications on telecom networks are currently embryonic. Its endorsement by TINA will allow it to gather pace as it will encourage network operators to experiment with this technology. CORBA (see Chapter 4) is a powerful distribution technology and has the potential to function as a unifying abstraction layer, presenting a homogeneous view of the transport capacity to layers above. In this way, it can contribute significantly to TINA's aspiration of accelerating the convergence of telecommunications and distributed computing technologies. Although TINA's implementations are still at a prototypical phase, they nevertheless give a clear indication as to the direction in which telecommunications are heading. In contrast to the Internet and computer networks, which are moving to more structured forms, telecommunications seek to 'flatten', or perhaps more appropriately, disentangle the impressive edifice of protocols, control and management mechanisms and transport architectures they have built over the years.

## 2.9    Future Trends and Enabling Technologies

It has been shown that moves can be identified in both telecommunication and computer networks that seek to pursue complementary trade-offs. In telecommunications, vertical integration and accruing performance gains are substituted for modularity and a stricter layering, while in data networks, layering rigor is substituted for an infrastructure that is more service-adaptive. While there are other conflicting moves as well, these are by far the most prevalent, and have thus vindicated the notion that telecoms and datacoms (along, perhaps, with broadcasting) are converging. Figure 2.1 attempts to graphically depict this process. Networks are systematic arrangements of a multitude of elementary constituents. Many attributes and properties can be ascribed to them, no specific set of which could fully identify them. For many different sets, however, computer and telecommunication networks will be found to converge. Figure 2.1 reports the dimensions that have been chosen for this specific convergence trail. Although it is certainly not drawn to scale and axes represent hardly quantifiable concepts, we believe it succeeds in capturing the trade-offs involved. It also demonstrates the conflicting aims and motivations underpinning contemporary technical innovations in both arenas. It can also be seen that active networks, whether employed on the Internet or in telecommunications, appear to belong to a class of their own quite distant in terms of the depicted plane's co-ordinates from both the present day Internet and telecommunication networks. Active Networks (see Chapter 6) represent an innovative way to structure an architecture and circumvent many of the trade-offs associated with more typical implementations. Active Networks have many different facets and schools of thought, and it may not always be appropriate to distinguish between Internet active networks and telecom ones. That's way these appear clouded in Figure 2.1.

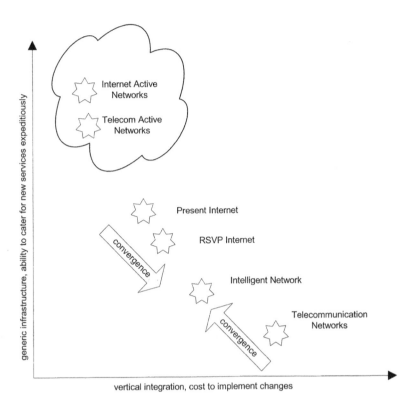

**Figure 2.1** Shifting trends in telecoms and datacoms

This book is concerned with telecommunication networks. It is of more importance, therefore, to examine the ways in which these networks can be adapted towards the objectives stated earlier.

We identify five enabling technologies: Object Orientation, Distributed Object Technology, Machine Independent Code and Mobile Agent Technologies. It is easy to see the pertinence of each of these.

The *Specification and Description Language* (SDL) (ITU-T, 1993) has been used (and is still in use) as one of the most widespread specification techniques for telecommunication system implementation. Telecommunication software has certain characteristics of its own, and for this reason merited a methodology and a formal specification language tailored to its own needs. Although technically, object oriented features were not included in the language until 1992, even before that date, the description of the language led to telecom software implementations which were object-based in nature, although not explicitly described as such. Following the formalization of the language in object oriented terms, and the adoption of the *Object Modeling Technique* (OMT) object model, telecommunication software will continue, to a steadily increasing extent, to be specified and implemented along object oriented lines, just as software for computer networks does. The importance of this convergence of methodologies or development paradigms may not be immediately obvious, since it is of a less concrete nature, but it can provide the permissive environment into which ideas, concepts and innovations from both domains can be freely exchanged and more readily applied. Chapter 3 describes what object orientation is about and how telecommunications can benefit from it.

*Distributed Object Technology* (DOT) is a middleware technology that disentangles application software from a transport network's conventions and complexity. Supporting the illusion of a flat communications space, it can be instrumental in promoting a consolidation of the Internet's and telecommunication network's infrastructure, at least from the transport layer and above. Central to DOT are the so-called 'location transparencies' it introduces, which are elaborated on in Chapter 4.

Machine Independent Code can be an answer to the extreme diversification of operating systems, hardware architectures and legacy equipment that characterize the contemporary telecommunications landscape. While usually associated with a mild performance degradation, its benefits can in certain cases outweigh performance considerations. Machine Independent Code, though a concept long formulated, remained largely latent until the advent of Java. Java and Machine Independent Code in general are examined in Chapter 5.

*Mobile Agent Technology* (MAT) is, like DOT, a middleware technology that works on the success of Java to realize on old idea: that of dispatching a piece of software to be executed remotely, local to the resources it needs to access (instead of having it executed locally and bearing the cost of remote access to the required resources). MAT can very naturally be coupled with DOT to attempt to offset some of the performance-related adverse effects of location transparency. MAT can be a very appropriate solution for certain telecommunication domains that have easily identifiable and anticipated patterns of interactions taking place among remote entities or processes. This is usually the norm, as we have noted in telecommunication environments. The Intelligent Network is a suitable example. Chapter 6 deals with MAT. The Intelligent Network concept is examined in more detail in Chapter 7.

It is useful to note that all of these technologies have emerged from computer networks, and it is there where they enjoy greater acceptance. Nevertheless, taken individually or even better by exploiting their complementarities, they can be very helpful in progressively enhancing telecommunication networks' potential to evolve. The first major landmark here is the Distributed Intelligent Networks, examined in detail as a case study of all those technologies in action in Part III of this book.

## References

ITU-T (former CCITT) (1993) Recommendation Z.100: Specification and Description Language SDL.

Meszaros, G. (1999) Design Patterns in Telecommunications Systems Architecture. *IEEE Communications Magazine*, **37**(4).

Saltzer, J. H. Et al. (1984) End-to-end arguments in system design. *ACM Transactions in Computer Systems*, **2**, 277-288.

# Part II

## Enabling Software Technologies

# 3

# Object Oriented Design Methodologies

## 3.1 Introduction

There are two fundamentally different ways to view software. According to the first, a program is viewed as a complex graph of data transformations. Information and data flow into the system and out of it, and at certain nodes are subject to modifications, transformations and all sorts of processing in general. We can term this approach the data centric one. In it, the basic constituents of a program are the data that is processed and the types of transformations that apply to it, or the sort of actions (behavior) that are triggered by certain data values.

The second approach views a program as a set of interacting entities, each performing a specific part. Entities exchange information by sending messages to each other. We will use the term 'object oriented' to describe this view of software. Here the basic constituents of a program are the objects that compose it. While data and its transformations are conceptually discrete entities in a data centric view, objects comprise both data structures and behavior.

The first approach emphasizes identifying appropriate information flows through the system, the kinds of transformations (functions) that should apply to these and arranging flows and functions to yield a coherent whole. This view of software stresses choosing appropriate data structures that will be used to represent information at each point inside the system. It also underlines the employment of appropriate algorithms and the arrangement of functions that will operate on the data structures of choice. Process oriented, function oriented or structured programming techniques are derived from such views of software.

The second approach emphasizes not information flows or their processing, but rather the allocation of responsibilities between independent entities that will mainly operate on their 'own' information, and will share information with others only when necessary. These entities are called objects, and Object Oriented (OO) techniques are founded on this notion of programming. In fact these techniques view intense information passing as an evidence of poor design. Objects should be so defined as to be for the most part independent and

*Object Oriented Software Technologies in Telecommunications,* Edited by I. Venieris, F. Zizza and T. Magedanz
© 2000 John Wiley & Sons Ltd

self-contained. We use the term 'interface' to collectively refer to the messages an object allows others to send to it. Later in this chapter, we note that this term has also a more technical, narrow sense, but until then we will use it informally to describe the set of messages an object can process. A fundamental tenet of *Object Oriented Programming* (OOP) is that interfaces should enhance the view of an object as an autonomous entity. That is, they should not correspond to overly specific instructions, prescribing 'how' to perform a certain task. An overly specific interface would indicate that an object is dependant on another. This might in turn mean that the object decomposition of choice was not optimal. Rather, interfaces should serve to delegate responsibility to the receiving object in order to perform a task by telling it 'what to do', instead of 'how to'. In this way, they will allow it more latitude and a greater flexibility in choosing the more appropriate data structures and algorithms for the responsibilities that have been assigned to it. Having 'autonomous' or 'self-contained' objects that interact on a higher semantic level by delegating responsibility is beneficial for another reason. It reduces the need for data sharing or excessive information passing. Imagine an object, say A, that will have a graphical appearance on screen. At certain points in time another object, say B, may need to instruct it to render some information on screen. To do that, B will have to use the interface A has made available. What kind of interface should that be? According to our discussion, an interface corresponding to 'present your information' would be suitable. An interface analogous to 'present your data item X in blue color and Times Roman font, 15 points, on black background and data item Y in red color and Courier new font' wouldn't. The first interface assumes nothing about the structuring of data that compose A's information. The second is aware of the existence of items X and Y and prescribes rendering that may be infeasible for a monochrome monitor or if the specified fonts are not installed. In the first case, if object A's internal representation of information changes, or if object A is used on a system with very rudimentary presentation capabilities, A's interface to B will remain the same and B's implementation will not need to change. Not so if the second interface is used.

The recognition that assuming little about an object's internal workings can safeguard against the need for extensive changes in other 'neighboring' objects, once these workings may change, has given rise to the concept of encapsulation or information hiding. Interfaces, objects sending messages to each other, and information hiding all underline the autonomous view of an object as a self-contained black box. This notion is so strong that it is not uncommon in object oriented literature to talk of objects 'assuming roles'.

Object oriented programming is appealing for a variety of reasons. Some are quite technical, others less so. This chapter focuses on object oriented programming and related methodologies, and its importance in the context of telecommunications. Accordingly, we examine object orientation and common object oriented methodologies, and attempt to explain their popularity and their pertinence for telecommunication software by giving concrete examples. A single chapter devoted to object orientation can only provide the most cursory examination. There are literally hundreds of books on this subject, of which the following three are classic (Booch, 1994;Rumbaugh, 1991;Jacobson, 1994).

## 3.2    General Principles of Object Orientation

A computer program usually attempts to solve a real world problem. The language in which the program is coded to a large extent determines the expressive power of the program, and formulates the abstractions and models the program will use to provide the mapping between the *problem domain* and the *solution domain*. For this reason, the evolution of programming languages had a far reaching effect on the evolution of programming techniques and methodologies.

In the early years of computing there were only the assembly languages, which provided a small abstraction of the underlying machine. Assembly language programmers did not have to remember the binary value of each command of the processor they were using. Instead, they had to remember symbolic names (mnemonics) which were more readable by humans. However, readability and some conveniences in using variables were almost where the list of advantages ended. There was still a very concrete, one-to-one correspondence between the machine (the machine language) and the assembly language. While machine languages consisted entirely of numbers, assembly languages used names. They had the same structure and set of commands, however. Each machine had its own machine language and correspondingly its own assembly language. As a result, a program written in assembly language was greatly dependant on the underlying machine. Porting of the program to another machine would require rewriting most parts of the program. The second generation of software tools appeared in the mid-sixties when the first *procedural* computer languages were born. These languages (such as Fortran, Pascal, and C) represented essentially a higher level of abstraction over assembly languages. Programmers did not have to know the internal architecture of the machine they programmed and, ideally, porting the program to another machine required only a simple recompilation. However, the main drawback of these languages, and in general of the procedural method of programming, was (and is) that still not a sufficiently high level of abstraction is provided. Programmers have to think in terms of the structure of the computer rather than the structure of the problem they are trying to solve. In the procedural method of programming, the programmer must establish an association between the machine and the real world. The effort required to perform this mapping, and the fact that it is irrelevant to the programming language, produces programs that are difficult to write and expensive to maintain. It must be acknowledged, however, that for certain application fields, a functional breakdown of a system might be more appropriate, and that using a procedural language might produce a more natural solution.

The object oriented approach, on the other hand, allows for a much better abstraction of a real-world problem. In addition to what we have noted in the introductory section, another major difference between procedural languages and object oriented ones is that, in the former, the association between the data and the functions that operate on them is only vaguely (and very often ambiguously) implied in the actual program, while the latter have formal mechanisms to introduce associations between functions and data. The object oriented approach goes a step further by providing tools for the designer to represent elements of the real world as basic or composite types in his model. This representation is general enough so that, once derived by a designer, it is not constrained to any particular set of requirements. As requirements evolve, the decomposition of real-world elements and concepts into objects will retain much of its relevance. In the object oriented world, we refer to these elements and their representations as 'objects'. The key idea is that the program is allowed to adapt itself to the real problem by adding new types of objects when appropriate. The expressive power of an object oriented language is enhanced so that when someone reads the code he is also reading words describing the solution. This is a more flexible and powerful mechanism than what we've had before.

In essence an object is the data coupled with functions that operate on it. Functions associated with an object are usually termed *methods* to distinguish them from the more mundane *global* functions which are not tied to an object's implementation. Couplings between data and functions that operate on them can also be shown in procedural languages. In this sense, 'conceptual' objects can also be identified there; the difference is that the language syntax does not provide a way to make this conceptualization explicit. So, there is such as a thing as using object oriented practices with procedural languages, but it has limited applicability. Moreover, even when it is applied, it rests on the programmer's ingenuity and 'clever hacks', and is not supported by development tools. On the contrary, object oriented languages are invaluable in describing associations between data and functions in a formal manner, which is comprehensible by the compiler and in turn offer a variety of other features like inheritance or polymorphism that can have no counterpart in a procedural program. By introducing the concept of an object, object oriented languages extend the notion of a *type* in procedural languages to include not only data structures, but also the types of functions that can apply to those structures.

Interfaces, inheritance, polymorphism; all these concepts have real world counterparts. The concept of inheritance, for instance, corresponds to the commonplace interpretation of the phrase 'an engineer is a person' when it is taken to mean that an engineer possesses all of a person's properties plus some more. Interfaces have a counterpart in the different facets any real-world entity can have when examined for different purposes. The engineer is a *taxpayer* for the IRC, while an *employee* for his or her company. Polymorphism finds a real world parallel in the adoption of uniform procedures at the delegation of the same task to different people, while the delegating entity is cognizant to (and perhaps exploiting the) fact that each will handle it in a different way. The mechanisms of object orientation are very anthropomorphic, and so allow us to describe the solution in terms of the problem, rather than in terms of the computer. In fact, there's still a connection back to the computer. Each object can be considered as a little machine; it has a state, and it has operations through which the object can receive and send messages.

In the object oriented world, any element of a problem one is trying to solve (e.g. an ATM switch, a physical link, etc.) can be represented as an object in his/her design. Therefore, any piece of software which is written in an object oriented language is actually a set of objects interacting by sending messages. The word 'message' here should not be confused with that of a packet containing protocol data. Messages in programming languages correspond to calling functions or an object's methods. However, at an abstract level, a design need not take that into account. The generic concept of messages is sufficient. At implementation time, when decisions are taken messages are almost invariably implemented as function or method calls, although nothing precludes other means to implement them. A traditional protocol-based approach may be sufficient if objects reside on different machines. In Chapter 4 it will be shown that distributed processing environments can extend the notion of 'method call' to include calls made across machine boundaries in a blocking or non-blocking manner.

Each object has a specific type which describes what kind of messages the particular object can receive. Obviously, all objects of the same type (type and class will be used interchangeably in this chapter) can receive exactly the same kind of messages. In other words, all objects of the same type have the same interface. The distinction between an object and its type (its class) is analogous to that between a variable and its type. The main effort in writing an object oriented program lies in defining the appropriate classes. Once a class (or type) has been defined, objects of it can be instantiated using a simple syntax.

We have noted that one reason why object oriented programming is so appealing is that it allows human programmers to think and visualize with concepts and objects they are familiar with. This is not simply a matter of individual preferences, as it can have a large impact on the maintainability and flexibility of the code. Rumbaugh (1991) notes that an attempt to reach a solution through decomposing a system's functionality may appear as the more straightforward way to implement a desired goal. It is, however, bound to be fragile as once the requirements change, the system may need massive reconstructing. For instance, data structures formerly built may no longer be appropriate. Algorithms may need to change. Small changes will propagate through the entire system as procedural languages do not offer the convenient abstraction of an object as a self-contained entity, and do not allow information hiding or encapsulation to be used to their full extent. In contrast, an object oriented solution is more likely to hold up as requirements evolve over time, since it is based on conceptualizations of the actual application domain itself, and the latter are less volatile than the *ad hoc* requirements of a single problem.

## 3.2.1    Interfaces

A suitable way for understanding the concept of an interface is the OSI 7-Layer Model, depicted in Figure 3.1. In this model, each protocol consists of a set of rules and conventions that describe how computer programs communicate across machine boundaries. The format, contents and meaning of all messages are defined. The 7-layer OSI protocol suite specifies agreements at a variety of abstraction levels, ranging from the details of bit transmission to high level application logic. A layer deals with a specific aspect of communication and uses the services of the next lower layer. Consequently, each layer has to know the services that are provided by the next lower layer. In other words, each layer must have a specific interface which describes the services that it can provide to the layer above.

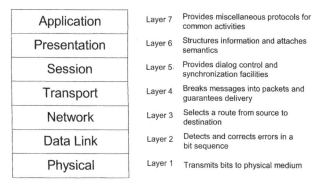

**Figure 3.1**   OSI reference model

In the object oriented world, each layer can be considered as a class, since each class has a specific interface that determines what kind of messages/requests the particular layer can receive (in reality, a layer is represented by more than one class). What other classes perceive is this interface in just the same manner as an OSI layer perceives only the services offered by its lower layer. The use of the word 'interface' here should not be confused with the more concrete notion of an *interface* as a programming language's construct (e.g. interfaces in Java). The notions of information hiding and encapsulation are pertinent in this context. The idea of information hiding is that an object should allow others to view it as a black box with an externally visible interface, while at the same time hiding implementation details of which only itself needs to be aware. Encapsulation is the mechanism used to implement information hiding. Usually, every object oriented language supports the so-called 'access specifiers'. These specifiers appear at the declaration of each element of a class interface and determine whether it is public, private or protected. Other more specialized specifiers may also appear, but these are the most common. The `public` specifier denotes that a certain element of the class is visible from all other classes. The `private` specifier restricts visibility to the object itself. Finally, `protected` requires an understanding of the mechanism of inheritance, and so must wait until Section 3.2.3. However, what are the elements upon which access specifiers are applied? For most programming languages they are *class* and *member variables* and *class* and *member methods*. Variables of an object are simply the data on which its methods (functions) operate. Class variables and methods apply to all objects of a class as a whole, while member variables and methods are associated with each object. Public, protected and private access specifiers can be used to define elaborate information hiding schemes that were not feasible in structured languages. By applying distinct specifiers to specific members of a class (e.g. a private access specifier for a certain variable and a public for another), information hiding can be refined beyond the gross scoping rules that were used in procedural languages. Restricting visibility of certain implementation details guarantees that external objects will not be able to depend upon them. Separation of concern will be guaranteed by the compiler, and not by the soundness of a design or the rigor of an implementation.

Up until now we have used (perhaps confusingly) the term interface, to refer to the portion of a class structure which is visible to other objects. We have done so with the intention of introducing a suitable analogy with the OSI reference model example presented above. The key word `interface` is actually used in several object oriented languages to describe a particular facet of a class. Since a class can have many facets (i.e. implement many interfaces), the term 'interface' as used in that context is more narrow than the use we have put it to so far. For the remainder of this chapter, unless otherwise indicated we will use the term 'interface' with its more technical, narrow sense.

The diagram shown in Figure 3.2 follows the format of the *Unified Modeling Language* (UML) (Fowler, 1999;UML notation guide, 1999;UML semantics, 1999). Each class is represented by a box, with the type's name in the top portion of the box, any member variables that we wish to describe in the middle portion of the box, and the member functions (the functions that belong to this object, which receive any messages sent to that object) in the bottom portion of the box. The '+' signs before the member functions or member variables indicates they are public (i.e. visible by external objects). Not public (i.e. private or protected) functions and variables are not visible by external objects in accordance with the objective of information hiding. Very often, only the name of the class and the public member functions are shown in UML design diagrams, and so the middle portion is not shown.

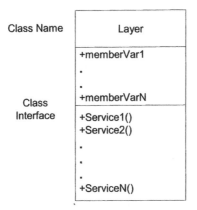

Figure 3.2   UML diagram for the abstract class `Layer`

Regarding the concept of interface as it used in its narrow sense, it can be noted that, in principle, a class can have as many different interfaces as the number of discrete subsets of its set of variables and methods. Figure 3.3 shows an (extreme) example of interfaces using UML notation for the 'implements interface' relationship. Not all interfaces that can be supported by this class are shown. Note also that two interfaces can be 'equivalent' and yet treated as different entities by most object oriented languages if declared under different names (e.g. IfC1 and IfC2).

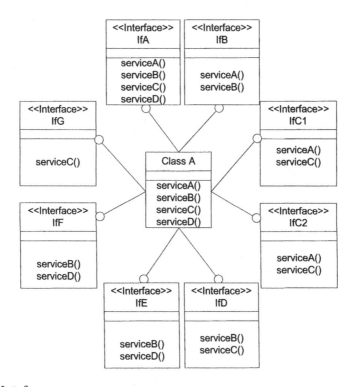

**Figure 3.3**   Interfaces

Figure 3.3 is a sort of caricature. One wouldn't expect to see such a UML diagram unless there were very specific reasons. For most normal cases it would be an indication of poor design. Classes usually support a small number of interfaces. Remember that the notion of a class implementing an interface corresponds to that of a class having a certain facet. When a class, for instance, implements two interfaces, say A and B, we expect it to be visible to some objects as being of type A while to others as of type B. It would be difficult to imagine an application where a certain object would need to use ten or twenty different facets.

Once a class is established, we can create as many instances (or objects) of that class as is required, and then manipulate those objects as if they are the elements that exist in the problem we are trying to solve. Indeed, one of the challenges of object oriented programming is to create a one-to-one mapping between the elements in the problem domain and objects in the solution domain. For instance, let us consider the above example with the network layers. An Internet router or an ATM switch may have only one instance of the Layer 3 class, but they usually have as many instances of the classes representing Layers 2 and 1 as the number of the physical ports they have.

## 3.2.2   Information hiding

Every program, irrespective of whether it is written in a procedural or an object oriented language, is a complex set of more elementary entities. In procedural code these entities are data structures and functions, while in an object oriented program they are objects. In hybrid languages such as C++ there can be both objects which combine data structures and functions, as well as 'stand-alone' (also called 'global') functions and data. The important thing to notice is that there are a lot of interrelations between all these elements. Dependencies abound, and it is common for a function's implementation to rest on a specific attribute of a data structure or for a data structure to include a specific field in order to be more easily handled by a function. More common and hazardous, however, is the situation where a certain data structure is accessible by more than one functions which can potentially modify its value. Since requirements and designs change, managing this mesh of dependencies is a daunting task. Ideally, such dependencies should be kept to a minimum. If not, ripple effects can be manifested when a small change takes place.

The principle of information hiding refers exactly to this process of hiding details of an object or function. It is clear that the developer can try to minimize dependencies when designing a program. However, effort by the developer is not enough. For instance, the original programmer can effectively eradicate dependencies between two functions by ensuring that they use separate data structures so their operations do not interfere with one another. Successive maintenance programmers, when confronted with a bug or in need of refurbishing the source code, will not be able to rest on the original programmer's assurance (even if one exists) that the two functions are unrelated. They will have to go through the code line by line to verify this. The moral is that although an individual programmer can skilfully reduce such dependencies, this will be of little use to successive programmers. Not to mention that the programmer himself might fail to guard against undesirable dependencies if left without any help from the programming language. What is needed are mechanisms that can impose this segregation between different functions or data structures at the language level. In this way, the compiler will restrict the visibility of certain program elements, and will ensure that the information hiding principle holds. Information hiding mechanisms appear in procedural languages, but are more advanced and complete in object oriented ones.

In a procedural language many important data structures exist at a 'global' level where they can be accessed by all parts of a program. The same holds for the so-called 'global' functions. They can be called by all parts of a program. Nevertheless, some sort of information hiding is supported by using a concept of *scoping* that limits the visibility of certain data structures defined locally to a function or functions defined inside another function (when the language permits). Scoping rules by themselves can do little to pursue real information hiding objectives, however.

Since object oriented languages group data structures together with functions, they are able to define stricter associations between them that in turn limit their visibility. In object oriented programming, and especially in 'purer than C++' object oriented languages like Java, it is usual for all data structures of an object to be declared private (meaning that they are not visible to other objects) and only for some methods to be declared public (corresponding to the class interface under the broad view of the term).

Encapsulation is a process closely associated with information hiding. In encapsulation the creation of larger entities out of combining more elementary ones is sought. This by itself facilitates information hiding, since it introduces more structuring and allows boundaries between different conceptual entities to become more tangible. Once boundaries are expressed using language syntax, they become language-level elements, and information hiding can more easily be applied to them. For instance, in procedural languages procedures serve to aggregate series of computer instructions, and so represent a form of encapsulation. Information hiding is also easier to apply as scoping rules can be defined restricting the visibility of data structures defined locally to a procedure. In the same way, objects group together data structures (possibly including other objects) and functions. They are also a form of encapsulation. Again, taking the definition of an object as a distinctive landmark in the source code's landscape, information hiding can be applied by specifying different visibility classes depending on whether the observing entity in question lies within the object's definition or not.

## 3.2.3   *Inheritance*

By itself, the idea of an object is a convenient tool. It allows system designers to package data and functionality together *by concept*, so they can represent an appropriate problem-space idea rather than being forced to use the idioms of the underlying machine.

However, there are many cases where two or more components of the system we are trying to model share some common characteristics. For instance, although 'Doctors' and 'Engineers' may represent two distinct classes of people, they still both belong to the more general class 'Human'. Therefore, it seems unreasonable to go to all the trouble to create a class and then be forced to create a new one that might have similar functionality. It would be more productive if one could take the existing class, clone it, and make additions and modifications to the clone. This is effectively what *inheritance* offers, with the exception that if the original class (called the *base* or *super* or *parent* class) is changed, the modified 'clone' (called the *derived* or *inherited* or *sub* or *child* class) also reflects those changes.

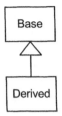

**Figure 3.4** In UML inheritance is denoted with an arrow pointing from the derived class to the base class

A type does more than describing the constraints on a set of objects; it also has a relationship with other types. Two types can have characteristics and behavior in common, but one type may contain more characteristics than another, and may also handle more messages (or handle them differently). Inheritance expresses this similarity between types with the concept of base types and derived types. A base type contains all the characteristics and behavior that are shared among the types derived from it. We create a base type to represent the core of our ideas about some objects in our system. From the base type, we derive other types to express the different ways in which that core can be realized.

In our example, each layer is represented by a different class with a different interface and functionality. However, all layers have some common characteristics that could be modeled in one base class. For instance, each layer should provide a way for receiving and sending data to the above layer, or each layer may need to be registered to the operating system during the start up procedure. Consequently, all these common characteristics can be modeled in a base class, and every time we wish to create a new layer we should derive our class from that base class. By doing so, all capabilities and features of the base class are automatically available to the new extended class. Moreover, if one of these common characteristics needs to be changed (e.g. how a layer is registered to the operating system), we will make the required modifications in the base class, and all changes will also be automatically reflected on the derived classes. In the non-object oriented approach, such a modification may require changes in the source code of all layers. Moreover, some of the layers may have more common characteristics which do not appear in other layers. In this case, we will define a new base class that will handle these common characteristics. The new base class will also be a child of the base class of all layers. For instance, there are some layers that have to check whether a Protocol Data Unit has errors or not. This extra common feature can be put in the new base class. Figure 3.5 illustrates this concept. All concrete layers (`PhysicalLayer`, `DataLinkLayer`, `TransportLayer`) ultimately inherit from the `Layer` class which contains only the methods common to them all. However, since both `DataLink` and `TransportLayer` include functionality for testing the packet checksum, a new class is derived from `Layer`, called `DetectErrorsLayer`. `DataLinkLayer` and `TransportLayer` then inherit from the `DetectErrorsLayer` instead of inheriting directly from `Layer`. Naturally this process, if left unchecked, can lead to a proliferation of types which can have adverse effects on the maintainability of the code.

Here is also the place to explain the protected access specifier. A member of a class specified as protected is not visible to an external object, just like a private. The difference is that, although private members are also not visible from descendant classes, protected members are. With regard to inheritance, all members that are put in a class and are expected to be used by derived classes as well are declared protected. Returning to Figure 3.5, `ReceiveDataPreviousLayer`, `ReceiveDataNextLayer` and `Register` methods in the class `Layer`, as well as the `CheckPacketCRC` method in the class `DectectErrorsLayer`, should be protected. The '#' sign is used to denote the 'protected' access specifier.

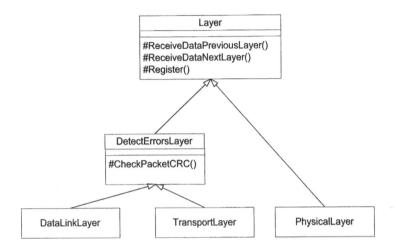

**Figure 3.5**  A simple example of inheritance in our layered model

When we inherit from an existing type, we create a new type. This new type contains not only all the members of the existing type (although the private ones are hidden away and are inaccessible), but more importantly, it duplicates the interface of the base class. That is, all the messages we can send to objects of the base class can also be sent to objects of the derived class. Since we know the type of a class by the messages we can send to it, this means that the derived class *is the same type* as the base class. In the above example, it can be said that the `DataLink` is a `DetectErrorsLayer` which is a `Layer`. This type equivalence via inheritance is one of the fundamental concepts of object oriented programming.

If we simply inherit a class and do not make any modifications or simply add a new method, the methods from the base class interface come right along into the derived class. This means that objects of the derived class have not only the same type, but also the same behavior. If a derived class was exactly similar to the base class, that wouldn't be particularly interesting.

There are two ways to differentiate a new derived class from the original base class. The first is quite straightforward: we simply add new functions to the derived class. These new functions are not part of the base class interface. This means that the base class simply doesn't do as much as the task at hand would require, so more functionality has to be added. In the above example, we created the `DetectErrorsLayer` class by inheriting from the `Layer` class and then adding a new method. This simple use of inheritance is, at times, the solution to many problems. However, we should look closely for the possibility that our base class might also need these additional functions. This process of iterative refinement is not uncommon in object oriented design.

The second way to differentiate a new class is to change the behavior of an existing base class function. This is referred to as *overriding* that function. To override a function, we simply create a new definition for the function in the derived class. By doing so, we keep the same interface as the base class, but we change the way that objects of the new class handle these messages. In our example above, all layers have the same interface on how they are registered to the operating system. However, the activities they undertake during their registration may differ.

## 3.2.4   Polymorphism

In the discussion of interfaces we noted that a class can implement many interfaces (i.e. support many facets). This is equivalent to saying that objects of this class can have many types. Multiple inheritance, when supported by a language, also allows objects to have many types. The question is: are we constrained to treat an object as though it had the most specific or alternatively the least specific type? The answer is no. If an object has many types, the programmer is given great latitude in choosing the most appropriate type for each task the object is put to use. This allows one to write code that doesn't depend upon specific types. As an example, consider the case of an Internet router. An Internet router has more than one physical interface which may be attached to different types of networks (e.g. one interface may be attached to a local Ethernet network, while another may connect the router to a Wide Area Network via ATM). Each physical interface has an associated network driver, which is, in our case, represented by the `DataLinkLayer`. The target is to write code for the third layer (i.e. the Internet Protocol layer) that will not depend upon any specific physical network interface. Again, the idea is to define a new base type that will include all the common characteristics of the physical level interface. Moreover, the design of layer 3 must be done in such a way that will use only the common characteristics of the base class, and nothing more.

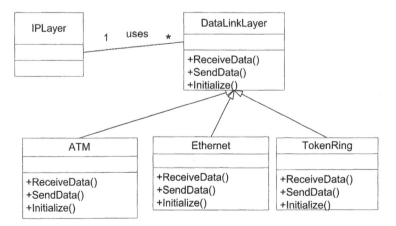

**Figure 3.6** Typical use of polymorphism where the design of the `IPLayer` class doesn't depend upon specific data link technology

In the UML diagram seen in Figure 3.6, the `IPLayer` has an association with the generic class `DataLinkLayer`. The `IPLayer` handles all physical ports as being of the same type without worrying about the peculiarities of each port. For instance, when the `IPLayer` wants to send a data packet to one port, it simply sends the message `SendData` to this port. How the data packet will be transmitted to the physical medium is something which is completely taken care of by the specific derived class. For ATM, the data packet will be split into ATM cells, while Ethernet will add the appropriate Medium Access Control (MAC) header. However, both ATM and Ethernet classes have the same interface on how they receive data from upper layer, that is, the `SendData` message.

Such a design is unaffected by the addition of new types, and adding new types is the most common way to extend an object oriented program to handle new situations. For example, we can derive a new subtype of a physical interface called `ADSL` without modifying the code in the `IPLayer` class. This ability to extend a program easily by deriving new subtypes is important because it greatly improves designs while reducing the cost of software maintenance.

## 3.3    Object Oriented Methodologies

The previous sections provided a short introduction to the concepts of object oriented programming. They described the concepts that are at the programmer's disposal, and it showed how these can improve reusability, modifiability, and extensibility, and enable a programmer to reach a more 'natural' design, that is, one where the concepts of the real world problem are more straightforwardly mapped onto the elements of a programming language.

What was not covered, however, is the process by which a designer moves from a statement of the problem to a detailed design that is then materialized using the aforesaid concepts. A systematic description of such a process is often called a 'methodology'. Methodologies vary considerably, and in most cases use divergent terminologies and emphasize different aspects of the software development process. They are also usually characterized by distinct graphical notations. These graphical notations are used to describe software designs in a language-independent manner. The advent of UML, which is a modeling notation, has brought forth some homogeneity, at least at the notation level. All methodologies tackle the problem through a series of steps, whereby at each step different kinds of concerns are prevalent and different sets of decisions are made. A clear-cut classification of methodologies would ask whether they are suitable for OO development or are more oriented towards structured programming techniques. Rumbaugh (1991) makes the case that a methodology for software development should not depend upon whether the implementation language of choice is object oriented or not. Instead, methods are presented that allow object oriented designs to be translated into structured languages and even object oriented databases. While this is not a book on methodologies, we provide a description of the Object Modeling Technique (OMT) methodology. This methodology uses a set of notations (the object model notation, the dynamic model notation, and the functional model notation) that share many similarities with UML. For a methodology that uses the UML notation see Jacobson (1999).

The OMT methodology has the following stages:

1.  *Analysis.* This step corresponds to what in other methodologies is described as the *Requirements Capture* phase. Information is gathered on what is the functionality that needs to be performed by the system. A model of the system is then constructed that encompasses this information. Attention should be paid to ensure that the model constructed in that step describes only the behavior of the system, and does not contain any assumptions on how the system will be implemented. That is, the system is examined from a 'use-case view', to borrow the terminology used by UML.

2.  *System Design.* In this step, high level decisions on the overall architecture of the system are made. Here is also the step where strategic decisions are taken. Such decisions can, for instance, include whether to structure the system into a large number of loosely coupled components, or into a smaller number of more tightly coupled ones. Decisions regarding high level non-functional optimizations (e.g. performance) and associated trade-offs also belong to this stage.

3.  *Object Design.* Taking as input the analysis model, the designer refines it further so that it includes not only the externally visible behavior, but also incorporates implementation decisions. These include algorithms, data structures and associations between objects. Decisions are made in this step taking into consideration practical needs (such as minimizing execution time, and minimizing required storage). Because of that, the analysis object model may be refined to include forms that are not strictly necessary from a conceptual point of view.

4.  *Implementation.* Finally, once a detailed object design has been reached, it should be relatively easy to translate it into a concrete programming language. At that final part, more objects may be added that were not identified at the analysis or design phases. They usually correspond to auxiliary data structures that have no counterpart in the real world.

OMT uses three types of notation to describe a system, namely the *object*, the *dynamic* and the *functional* model. Each notation is applicable at all phases of the development process. The object model is used to capture the static structure of the system. It contains information about relationships between objects (to which the OMT pays great attention), as well as the inheritance graphs of the system. The dynamic model is used to specify the dynamic behavior of the system, and is the model where control aspects are best grasped. It includes state diagrams that contain the states of the system and depict transitions among them. Finally, there is the functional model that emphasizes data flows and shows the transformations that data is undergoing as it 'flows' through the system.

The description of the OMT methodology concludes the introductory part of this chapter. The aim was to give the reader an understanding of what OOP is about and an appreciation of the complex processes involved in developing real-world software. The following sections examine object oriented practices put into use for developing telecommunications software.

## 3.4    Object Oriented Approaches in Telecommunications Software

Telecommunication software has traditionally been implemented using procedural languages such as C. These languages have certain advantages compared to object oriented ones, that are highly valued in the telecommunications world. Object oriented languages frequently employ mechanisms such as polymorphism or virtual functions which incur significant overheads. Late binding, for instance, is implemented using an extra level of indirection for method calls. Statically compiled function calls can be more effective. Most OO languages represent a significantly higher level of abstraction than structured ones. This has the usually cited benefits of fostering increased programmers' productivity and enabling re-usability of software solutions. However, it is also more difficult to write code that controls low level machine functions in a high level language. From the compiler's point of view, it could be more difficult to generate efficient code for a higher level language than for a lower level one. Simply put, the abstraction gap the compiler has to narrow is much wider in the former case. Another explanation for the wide use of structured techniques in that context would take note of the stringent reliability demands that are prevalent in telecommunications. The more complex a language, the greater the potential for errors. Simple languages such as C usually offer one way to do things (i.e. they satisfy the criterion of orthogonality), and allow the developer fewer degrees of freedom. The risk of introducing bugs is accordingly reduced. Last but not least, real-time requirements (which abound in telecommunications) are best met when using a language that is 'closer to the machine'. These have been the main reasons that made procedural languages popular in the telecommunications world.

It can be observed that most of the reasons put forward do not qualify for inherent constraints of object oriented languages. However, they reflect current practicalities in commercial implementations, and accordingly the practices that standardization bodies have in mind when producing telecom specifications. The rapid development of telecommunication networks – both in terms of their capacity and their set of offered services –, as well as the ongoing convergence of broadcasting, telecoms and datacoms, have certain implications in the telecommunications software realm. More specifically, they make object oriented techniques more attractive and process oriented ones less so. In the following we examine the reasons for this.

First, the market is becoming more dynamic, meaning that time-to-market becomes a crucial factor and that the ability to leverage on already existing components can create a competitive edge. Secondly, the scale of most modern implementations is simply prohibitive to create everything from scratch. The skill base necessary to successfully complete a project is also wider, encompassing people from telecoms as well as from the *Information Technology* (IT) industries. Both of these factors underlie the importance of an effective reuse process. Object oriented practices and methodologies result in code that is far more likely to meet the reusability demands than is code produced with process oriented techniques. Similarly, specification, design and development times will be reduced. The systems will be more amenable to future modifications and extensions. Inheritance in particular is unparalleled in its ability to enable reuse of off-the-shelf or in-house developed libraries. Function oriented approaches are at best able to allow for 'source code reuse' (copy-pasting function and data structure definitions from one project to another) when modifications or extensions are needed. There is no comparison between these two methods of reuse: inheritance is provided by the language itself, is applied in an orderly and systematic manner, and is supported by the compiler and the run time type checking mechanisms. Source code reuse, on the other hand, bypasses language, compiler and run-time mechanisms and overall makes for a very poor substitute. A point here is that it is possible again to use a procedural language like C in an object oriented manner and exploit some of the advantages of object oriented techniques in terms of reusability and modifiability. This approach, however, may seem unnatural, requires skill and discipline on behalf of the designer, and in some cases may not be available at all (e.g. in the case of third-party libraries where the other programmers did not follow the same technique). The mechanisms of inheritance and polymorphism allow a developer to gradually add more methods and facilities in a class library, enabling the system to evolve gracefully. A functional breakdown of a system is not nearly as adept at evolving over time as requirements change and modifications become necessary.

It is for the reasons mentioned above that telecommunication projects using structured techniques are increasingly facing long development cycles and high costs. These techniques fail to offer the levels of reusability, modifiability and extensibility that would be necessary. Correspondingly, implementations that use these techniques do not scale. They also lack the flexibility and the sophistication that is often required to ensure the viability of newly introduced services. While (structured) service creation environments exist and are used in commercial deployments, they fall short of addressing the complexity of required service logic and the need to interwork with off-the-shelf object oriented frameworks. They are also characterized by a procedural development logic, and do not lend themselves naturally to object oriented paradigms. This is in most cases either imposed by relative standards or at least influenced by their 'aura'. For instance, the standard way to create services in an Intelligent Network is by interconnecting the so-called *Service Independent Building Blocks* (SIBs). The SIBs represent a breakdown of a system's functionality into very elementary facilities. These facilities are service-independent so they can be combined in many ways to produce the actual services. This method of creating services is procedural in nature.

Trying to derive an object oriented implementation from a specification written with a procedural implementation in mind can be hard, and the success of the outcome is conditional on the abstractness of the specification and the lack of constraining or 'paradigm-bound' assumptions. Efforts directed at promoting object oriented concepts at the specification level for telecommunication standards can have a subtle but more permanent long term effect. Section 3.5 reports two examples of telecommunications architectures that were from the beginning specified along object oriented lines. The remainder of this section discusses object oriented approaches in developing telecommunication software.

### 3.4.1    Implementation of the intelligent network service switching function

The Intelligent Network Capability Set 2 recommendations (ITU-T, 1997) describe an entity called the *Intelligent Network Switching State Model* (IN-SSM). This entity is found in the *Service Switching Function* (SSF), which is located in the *Service Switching Point* (SSP) in commercial IN implementations. To understand the importance of the IN-SSM we first have to consider its context. An SSP is simply an ATM switch that is equipped with more intelligence. It is able to recognize important events in the course of a call/connection and (upon detecting such an event) to request instructions from a *Service Control Point* (SCP) on how to proceed. ATM switches have a more or less straightforward functionality: they spend most of their (significant) processing power switching packets from one interface to another. Naturally this procedure involves inspecting internal tables to determine the Virtual Circuit Identifiers (VCIs) and Virtual Path Identifiers (VPIs) of the outgoing packets, and setting up such Virtual Circuits (VCs) or Virtual Paths (VPs). The IN concept emphasizes segregating  the 'logic' of a service from the mere switching functionality and localizing the former in dedicated servers (the SCPs). It follows then that the semantics of the protocol by which the SSPs request and receive instructions from the SCPs are more abstract than the concepts upon which the implementation of an ATM switch is based. This is where the IN-SSM comes into play: it is responsible for bridging that gap and offering to the SCP a high level view of the switching capabilities. It does so by employing object oriented abstractions of common switching and transmission resources. In essence, the state of a connection is at each time modeled using a set of objects (corresponding to parties, connections and legs) and the arbitrary associations developed among them. Figure 3.7 depicts a typical such set.

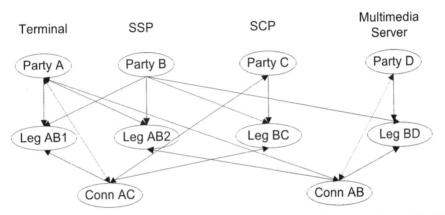

**Figure 3.7** Typical associations between IN-SSM objects in the case of an interactive multimedia retrieval IN service

Figure 3.7 corresponds to the case of a terminal (depicted in the IN-SSM as Party A), being connected to both an SCP (connection AC) and a Multimedia Server (connection AB). Both connections consist of two legs: the first connecting the terminal to the switch (the SSP); and the second connecting the switch to the SCP or the Multimedia Server.

The way in which the IN-SSM is specified in the relevant recommendations also calls for an object oriented implementation. An obvious approach would be to map each IN-SSM object into a class providing it with methods to connect it to other classes or retrieve a view of its state. For instance, a `Party` class could contain the following methods:

- `public boolean owns(Leg leg);`
- `public void attach(Leg leg);`

The first of these two methods queries a `Party` object to determine the ownership of a `Leg` object. The second attaches a `Party` to an already created `Leg`. The IN-SSM definition strongly hints for an object oriented implementation, and clearly indicates an appropriate object breakdown. It is conceivable that certain commercial SSP implementations must have resorted to function oriented techniques simply because they were created with a procedural language. Such constraints (and possibly extreme performance considerations) aside, an implementation in an object oriented language seems the most natural solution. Apart from the connection and transmission resource abstractions (legs, connections, parties), a number of *Finite State Machines* (FSMs) has to be maintained as well. This is because it is very common for switching equipment to have to maintain some sort of state information. The processing of each message it may receive is dependant on what messages have preceded or, alternatively, on what state the equipment is currently in. FSMs are a recurring theme in programming, and a lot of well known (and widely practiced) methods are available for their implementation. For an object oriented implementation, however, one particular method that should be mentioned is the FSM pattern (Gamma, 1995). The FSM design reported there rests heavily on inheritance and polymorphism to produce an implementation that can easily add new states when appropriate without incurring any modifications on existing states or transitions between them.

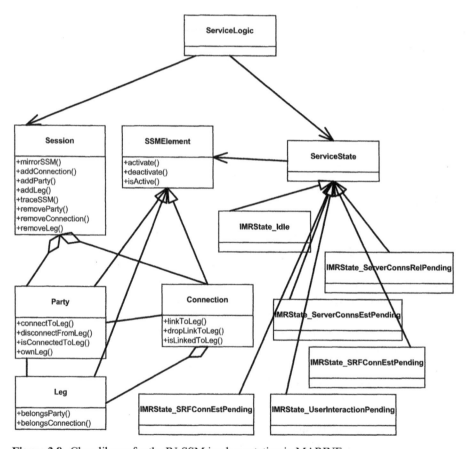

**Figure 3.8**  Class library for the IN-SSM implementation in MARINE

Figure 3.8 presents (in UML notation) the class library that was developed for the IN-SSM objects in the MARINE project. They were implemented using Java and were used in the IN-SSM mirror that has to be kept at the SCF. On the right side of the diagram the classes implementing the FSM pattern are shown (`ServiceState` and its subclasses). This library is presented in more detail in Chapter 9.

A description of the object oriented model for the SSF implementation at the ITALTEL switch (which was also used at that project) can be found in Venieris (1998).

### 3.4.2   Implementation of the intelligent network service logic programs

*Service Logic Programs* in commercial IN implementations are derived using a *Service Creation Environment* (SCE) which is a development environment offered as part of the platform. The service developer uses an intuitive *Graphical User Interface* (GUI) to interconnect building blocks that represent the SIBs described in the IN recommendations. In the general case, the tool using the information that is presented in the way SIBs are interconnected and parameterized is able to generate some sort of code. The kind of code that is generated depends upon the tool used. It could be the case that this code is only used to glue together already existing code fragments encompassed in the SIB implementation. In another scenario, the code may be interpreted by an execution machine designed to handle SIBs. Generally, however, a set of libraries (of classes or functions) is available, and the code produced by the tool is linked with these libraries. From the SCE development point of view, these libraries constitute a valuable asset. Well structured libraries can greatly improve the efficiency of the SCE since the latter can leverage on them and avoid producing non-service specific code that could also be part of a library. Moreover, SCE becomes less prone to changes in the environment, such as installation of a new communication protocol, since the necessary changes can be accommodated at the class library level, leaving the SCE intact. Chapter 9 presents such a class library that uses different inheritance hierarchies to separate concerns arising from different constituents of the environment. An organizational view of that library is presented in Figure 3.9.

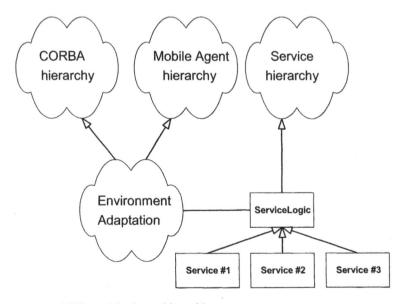

**Figure 3.9**  Use of different inheritance hierarchies to separate concerns

What is actually suggested in Figure 3.9 is that separate hierarchies are implemented for each 'implementation relevant' component of the environment. By 'implementation relevant' we mean containing assumptions or using facilities of the environment that will leave a footprint on its code. These facilities and components will require changes in the code if they themselves change. Such are those that call for specific arrangements, that are actively used, interacted with or in any way exploited. Assumptions implicitly made about the environment can often go unnoticed. In those cases, subtle dependencies are introduced that can affect the re-usability of a library in a different context (or even in a slightly modified one). Directly extending third-party class libraries to derive problem-specific implementations can in that sense be very dangerous, as:

- assumptions made by the class library vendor become ours as well, and
- these assumptions are not readily identifiable.

When possible, all third-party code and libraries should not be inherited from but interacted with through special interfaces developed for that purpose. These interfaces (depicted in the cloud 'Environment Adaptation') should represent an abstract view of the services the problem-specific code expects to obtain from its environment. They should represent both 'compile-time services' (e.g. facilities embodied in libraries to be linked at compile time) as well as 'run-time services' (more volatile services available at runtime). A typical third-party library is an example of the first category, while the existence of a CORBA Naming Service (and the means to interface with it) falls into the second. The 'Service hierarchy' cloud in Figure 3.9 represents a specifically developed hierarchy for the problem at hand which it is safe to inherit from. The clouds 'CORBA' and 'Mobile Agent hierarchy' are environment specific class libraries that do not pertain to the problem and are orthogonal to any solution. Thus, they are more loosely coupled with the code since they are interfaced and not inherited from.

It is not always possible to proceed with these guidelines. A typical example would be if it were necessary to use a framework to develop the code. Frameworks are more tightly coupled with the developed code than are libraries and can pose re-usability problems that are not so easy to circumvent. However, frameworks are more widespread in the IT industry. Mainstream telecommunication software is not developed using frameworks although this situation is likely to change. Another acknowledgement has to be made with respect to the fact that it is usually more productive to directly inherit from a library than to use all these elaborate (and performance impeding) interfacing schemes. Also, telecommunication environments are far less volatile than IT ones (reflecting the higher capital expenditures with which they are associated), so some of the concerns presented here are less worrying.

## 3.5    Network Management and Service Engineering

Methodologies for designing modern broadband architectures (or producing specifications for them) are no longer constrained to a strictly procedural approach. Methodologies of that kind were effective in designing systems like the *Public Switched Telephone Network* (PSTN). These systems implied or assumed a strong correlation between service (e.g. offering a voice circuit) and infrastructure. Such models are no longer adequate given the convergence of IT and telecommunications. As services are introduced, evolved or withdrawn at a more precipitous pace than in the past, modern telecom infrastructure is designed to be independent from the set of services that employ it. This independence is evident at all levels: service creation, service or network management, service or network control. Placing a telephone call was a conceptually straightforward procedure that can be viewed in a functional manner. All the more so, since it rested on a dedicated infrastructure. There are no longer dedicated infrastructures as such. Modern networks are more generic in that respect and are not bound to a specific 'service paradigm'. Merging of IT and telecommunication networks fosters and in fact entails merging of models, concepts and methodologies. This process is symmetrical on both constituents. Although the legacy of functional system specifications is still dominant, the telecom industry has appreciated the merits of the object oriented paradigm as a specification tool for network control and management. This has paved the way for its acceptance as an implementation tool as well. Distributed processing specifications of the IT industry, on the other hand, are also engineered in a manner that will allow them to retain their pertinence in the new environment. This is reflected in provisions for non-functional characteristics like real-time requirements, performance, guaranteed Quality of Service (QoS), robustness and security. Traditionally these have been primary concerns for the telecoms world, but have not always been at the focal point of attention when producing IT network specifications. Wishing to expand the range of applicability of IT products to include telecommunication markets, relevant specifications pay an increasing attention to performance, quality of service and real-time requirements. This is especially true for middleware or network protocols specifications, as these have the greatest potential to be employed in other domains. For instance, *Common Object Request Broker Architecture* (CORBA) specifications are augmented with real-time characteristics and real-time Object Request Brokers are available. This will encourage the use of CORBA, for instance, to replace Network to Node Interface (NNI) signaling. The next generation of the Internet suite of protocols will similarly be better equipped than their predecessors to guarantee a certain quality of service level.

The following subsections examine two telecommunication architectures that were for the most part defined along object oriented lines. They serve to demonstrate the use of object oriented approaches as specification and implementation tools in the telecoms world. CORBA, a promising distributed technology that is object oriented in conception, and the way it affects telecommunication software, is described in Chapter 4.

### 3.5.1   *Telecommunications Management Network*

The *Telecommunications Management Network* (TMN) is a telecommunications concept aiming to standardize interfaces and protocols for configuring, managing and monitoring telecommunications equipment. It pervades all aspects of management of telecommunication systems: from day-to-day micro-management of network equipment to the more strategic management of services. It caters for all five elements of the *Fault, Configuration, Accounting, Performance and Security* (FCAPS) management model. Key to its definition is an hierarchy of management functions that serves to separate concerns and encapsulate functions within individual layers.

The TMN distinguishes between two types of telecommunication elements: *Network Elements* (NE) that represent the managed systems, and *Operating Systems* (OS) that manage the former. The primary aim of TMN is to specify the interfaces that govern the associations between managing and managed objects, and to introduce new interconnection relationships (along with their interfaces) when they are applicable.

TMN adopts an object oriented approach whereby network resources (or clusters of them for that purpose) are modeled under the convenient abstraction of a single entity, a *managed object* to be accessed, monitor or manipulated by software assuming a managerial role through the OSI management service/protocol using the *Common Management Information System* (CMIS) and the *Common Management Information Protocol* (CMIP). This object oriented specification lends itself naturally to object oriented implementations that ensure re-usability of software solutions out of the context for which they were originally sought.

Furthermore, ITU-T has adopted OSI Management standards for use by the TMN. OSI Management specified in the ITU-T X.700 series will provide for TMN *Operations, Administration, Maintenance and Provisioning* (OAM&P) functionality. These functions are also modeled in OSI Management in an object oriented manner.

### 3.5.2   TINA

The *Telecommunications Information Networking Architecture* (TINA) defines an architecture with the ambitious goal of integrating the Internet and traditional telecommunication systems. It places a strong emphasis on providing versatile multimedia services over the Internet and traditional telecommunications networks while keeping the costs of creating, deploying and managing these services in check.

The philosophy espoused by TINA is a separation of the high level applications from the physical infrastructure. This isolates the more stable control and management aspects from underlying technologies. The control and management functions are integrated into a unified architecture which is supported by a single Distributed Processing Environment. Control and management functions and procedures are intended to be generic, so that only service-specific functionality needs to be developed for any new service.

TINA aims for increased levels of re-usability and extensibility to make services and architectures easier to develop, test and manage. This is natural as it has been shown that lack of extensibility and reusability can pose serious problems to developing complex distributed telecommunications software. Extensibility favors a piece-wise approach, and can allow a service to ship in time while ensuring that modifications and enhancements may follow without requiring scrapping an already deployed service. Reuse allows a service provider to leverage on a extended base of off-the-shelf or in-house developed components. Finally, the perplexities of a complex system should be abstracted and the dependencies incurred when implementing for a specific technology should be minimized. Both these objectives can expedite the service creation process and increase the applicability of developed components (since the latter are not bound to a specific infrastructure). Moreover, by doing so, they facilitate rapid application development using sophisticated service creation environments.

As has been noted, the main benefits of OOP revolve around extensibility and re-use, which are exactly the goals that TINA is pursuing. It is little wonder, therefore, that the specification relies strongly on OO concepts and methodologies. In fact, object oriented analysis and design is among the stated principles that govern TINA specifications. Chapman (1995) states that: 'The intention is to make use of recent advances in distributed computing (e.g. Open Distributed Processing (ODP) and Distributed Communication Environment (DCE)), and in object oriented analysis and design, to drastically improve interoperability, re-use of software and specifications, and flexible placement of software on computing platforms/nodes'.

Overall, the TINA architecture is based on four principles:

- Object oriented analysis and design
- Distribution
- Decoupling of software components
- Separation of concern.

It is not the objective of this subsection to provide a comprehensive coverage of TINA. Related specifications are both extended and at certain points insufficiently concrete. Chapter 4 provides a high level overview for the interested reader. What is important is to observe the affinity that common object oriented practices in the IT world have with these principles. Distribution of software components over different parts of the network is strongly dependant on the application of Distributed Object Technologies, and indeed, the TINA consortium has adopted CORBA for that purpose. Decoupling of software components coincides with the objectives of encapsulation and information-hiding. Separation of concern comes naturally once the above goals have been fulfilled.

Concluding, TINA outlines an 'ideal' architecture where IT and telecommunication networks coexist harmoniously, and where the introduction of new services is unimpeded as:

- Management and control functionality is generic.
- Underlying transport technology is abstracted.

Subsets of the TINA architecture have been realized in either commercial or prototypical implementations. From the point of view of this chapter, what is noteworthy is that object orientation was widely employed to specify a telecommunications architecture and to implement it.

# References

Booch, G. (1994) *Object-Oriented Analysis and Design With Applications*. Addison-Wesley Object Technology Series.

Chapman, M. (1995) *Overall Concepts and Principles of TINA*. Available from www.tinac.com

Fowler, M. and Scott, K. (1999) *UML Distilled, Second Edition: A Brief Guide to the Standard Object Modeling Language*. Addison-Wesley Object Technology Series.

Gamma, E. et al. (1995) *Design Patterns: Elements of Reusable Object-Oriented Software*. Addison-Wesley Professional Computing.

ITU-T (1997) The Q.122X series of recommendations (Intelligent Network Capability Set 2).

Jacobson, I. (1994) *Object-Oriented Software Engineering: A Use Case Driven Approach*. Addison-

Wesley Object Technology Series.

Jacobson, I., Booch, J. and Rumbaugh J. (1999) *Unified Software Development Process*. Addison-Wesley Object Technology Series.

Rumbaugh, J. et al. (1991) *Object Oriented Modeling and Design*. Prentice Hall.

UML Notation Guide. Available from http://www.rational.com

UML Semantics. Available from http://www.rational.com

Venieris, I.S. and Hussmann, H. eds. (1998) *Intelligent Broadband Networks*. John Wiley & Sons Ltd.

# 4

# Distributed Object Technology

In the last few years, advances in the Information Technology and Telecommunications fields have yielded an increase in computer performance along with a decrease in data transmission costs. Moreover, the availability of high speed, local or wide area networks and the proliferation of software applications and services has stimulated the realization of software applications exchanging data over a network or making active use of network resources. The complexity of these applications emphasized the need for flexibility and simplicity at the application level, and has as a result, paved the way for the introduction of distributed software in various fields of computing.

A distributed application is composed of a set of communicating processes which are in the general case executed on different processing nodes and are co-operating to solve a common problem. Whenever this set of processing nodes can be viewed by the application itself as a single node, the application is actually distributed. This means that there are no explicit references either in the client or in the server source code to the process's physical locations (IP addresses, TCP/UDP port numbers), as happens with traditional client-server applications. Naturally, very few applications have such references hardcoded in their source code; they usually read them from a disk, or the user may provide them and the program treats them as variables to be initialized at runtime. The point remains, however: such applications directly manipulate low level physical addresses. Irrespective of the scenario, ultimately, concrete transport network addresses have to be used; with distributed computing, middleware is installed on each node that allows the applications to use a higher level representation of the network's resources. In this scenario, the representations two communicating processes may hold for their peers are transparently mapped by the middleware into the lower level addresses. Such communicating processes are appropriately called 'distributed objects' or 'components'. The term 'distributed' hints at the fact that, since there is an intermediate step of mapping high level concepts to a lower protocol's functionality, the actual, physical location of the application may change without causing peer applications to malfunction as long as the middleware can take care of the mapping transparently. It is also used to indicate the effect the presence of the middleware is having; that of allowing processes located on different nodes, subnets or even networks to be viewed as residing on a single, unstructured, plane. The applications are then said to be operating 'on top of' a distributed processing environment, in the sense that they can be viewed as being located on this flat space.

*Object Oriented Software Technologies in Telecommunications,* Edited by I. Venieris, F. Zizza and T. Magedanz
© 2000 John Wiley & Sons Ltd

Note that the design and implementation of distributed object based applications follows the object oriented paradigm, which foresees the tight integration within a single object between the data and the behavior characterizing the target entity. The distributed approach is clearly in contrast with the centralized and traditional network-based approach, where system component locations are within the same machine or, if distributed, are well known and fixed at execution time. In fact, the main difference between a distributed and a network-based system is the concept of location transparency, which allows both the user and the distributed application itself to disregard component locations. Since applications can be thought of as being plugged on an unstructured plane that incorporates no topological information, it follows that at that level the actual locations of individual components are abstracted.

## 4.1    General Principles

This section introduces the general principles that form the basis of *Distributed Object Technology* (DOT). A comprehensive introduction to distributed computing, can be found in Orfali (1996).

As the basic constituents of any distributed object architecture one can identify:

- *Remote Method Invocation.* The remote method invocation mechanism is an evolution of the *Remote Procedure Call* (RPC) method for object-based distributed scenarios. While classical RPC foresees the invocation of a function, which is naturally separate from the data it handles, in an object-based distributed environment, method invocations are invocations of operations (methods) on specific object instances. So remote method invocation readily supports the object oriented programming model, and allows an object's methods to be invoked by remote as well as local code. In fact the invoked object has no way to distinguish between a local and a remote call unless, of course, a method's parameters are used to convey that information. Implicit in this approach is the requirement that the concept of a (memory) pointer to an object is extended to also include pointers to remotely located objects. Pointers to such objects that reside in different memory spaces are called 'object references'. Support for the object oriented model then comes as natural as invoking an object method through an object reference, and can encompass such features as method overloading or method overriding which allows for polymorphic calls to be made. For instance, different classes can implement the same method in different ways while a remote object can hold 'upcast' remote references to a base abstract class which contains that same method. The remote object can then treat the instances of the concrete subclasses invariably and remotely invoke polymorphic methods using the references it holds. Once these invocations have reached the address spaces of the actual objects, the invocation will be dispatched to the appropriate implementation, as foreseen by polymorphism.   In practice, support for object oriented concepts may not be that seamless. For instance, certain distributed object middlewares do not support method overloading.
- *Implementation Independent Interface Definition Language.* The object oriented paradigm foresees that objects are externally visible only through their public attributes and methods. In a distributed environment this means that an object is characterized by what can be described as a contractual interface. Such a contractual interface actually

determines which internal data and operations are available to clients. It corresponds to the notion of an object's interface or facet as used in Chapter 3. Since both client and server objects could be implemented in different programming languages, a neutral, declarative language which allows us to properly define object interfaces is used. Since that language is implementation neutral, mappings have to be specified between it and concrete programming languages.

- *Location Transparency.* Clients do not need to be aware of the location of the server components with which they interact. Moreover, they do not perceive the mechanism used to communicate their invocations to server objects or to activate them. In fact, all these aspects are handled by the middleware, often called the software bus, which is responsible for determining the object that implements the interface required by the client, discovering its location, activating it and supporting its communication with the client. To provide for these capabilities, the middleware is responsible for handling the different address spaces or protocols that may apply at each processing node.
- *Naming.* In a distributed system each object can be bound to a user-defined name within a given naming context or to an existing directory (e.g. X.500), and this name can be used to obtain the associated object reference. As has been explained, object references can then be used by clients to unambiguously refer to the server objects and invoke operations on them. In this scenario, the naming service handles the mapping between names that are unique within a naming context and object references, which are unique within a network. Server objects that want to be reachable register themselves with the naming service, so that any client can contact them. Clients are able to retrieve object references to distributed (remote) objects provided that they know the names with which they query the naming service.

## 4.2    Distributed Object Architectures

The need for distributed systems and related standards has driven both research and development activities, as well as standardization efforts. Essentially, the goal is the specification of a distributed object architecture that will, acting as middleware, make developing and managing distributed applications easier. The Open Software Foundation (OSF), Component Integration Laboratories, Object Management Group (OMG) and Microsoft have produced their own solutions, respectively called the *Distributed Computing Environment* (DCE), *OpenDoc, Common Object Request Broker Architecture* (CORBA) and *Distributed Component Object Model* (DCOM).

Currently, the two major competing standards are CORBA and DCOM. CORBA is the product of the OMG consortium. OMG started standardizing CORBA in around 1989, and now includes most main IT companies. It is considered the leading IT consortium and some of its standards are also recognized by the International Organization for Standardization (ISO). CORBA is an architecture based on the concept of a common software or object bus allowing for distributed object interoperability, and providing a wide set of services to interacting objects. The bus is called the *Object Request Broker* (ORB), and is distributed only in a conceptual sense. What each vendor actually produces is localized components of that ORB which are executed in a processing node. These components communicate using, by necessity, lower level mechanisms (e.g. sockets) in order to provide the illusion of a distributed bus. Originally all these localized ORB components (called ORBs themselves) had to be from the same vendor. It was not until version 2.0, when an interoperability format to be used between different vendors' ORBs was specified, that the object bus could encompass products of different vendors. The current version of the CORBA specifications is 2.3 (OMG CORBA 2.3, 1998), representing the third enhancement made to the 2.0 version. Standardization activities are still ongoing towards CORBA 3.0.

Microsoft started later, in 1994, by defining an alternative standard, called the *Component Object Model* (COM), originally for supporting distributed objects within a single machine. COM was thought of as an object bus for the *Object Linking and Embedding* (OLE) components. Later, to deal with real component distribution, Microsoft specified Distributed COM, which provided for location transparency, and allowed remote in addition to local communication.

This section focuses on these widely used distributed object architectures. The OMG's *Object Management Architecture* (OMA) and the more refined CORBA architecture are analyzed in detail. The Microsoft DCOM architecture is also introduced, and a comparison between the two approaches is given.

## 4.2.1   Object Management Group Architecture

This section starts with a general introduction to the OMG Object Management Architecture, proceeds with a detailed description of the CORBA components and services, and concludes with an overview of CORBA standardization efforts.

### 4.2.1.1   Object Management Architecture

The Object Management Architecture was defined by OMG in the early nineties with the goal of providing a high level specification of the functionality needed for object oriented distributed processing. Such functionality revolves around: how distributed objects should be defined and created; what communication mechanisms are required; and how methods should be invoked on remote objects.

The architecture comprises an object model and a reference model. The object model shows how objects must be described in the context of a heterogeneous distributed environment, while the reference model describes possible interactions between these objects.

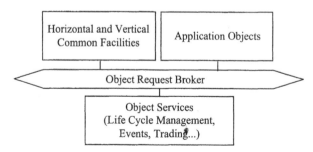

**Figure 4.1** The Object Management Architecture

In the OMA object model, an object is an encapsulated entity characterized by a specific identity able to provide external objects with services, which are accessible only through publicized interfaces without requiring knowledge of the object's specific implementation. Figure 4.1 shows the four main elements of the architecture: the *Object Request Broker*, the *Object Services*, the *Common Facilities* and the *Application Objects*.

The ORB, which corresponds to the CORBA object bus mentioned above, is the core of the OMA architecture. It provides the mechanisms to support object location transparency, server object activation and inter-object communication.

The Object Services (OMG Services, 1998) are a collection of domain independent low-level services, whose interfaces have been defined using the Interface Definition Language (IDL). These services extend the ORB functionality with fundamental services, like: lifecycle management, event dispatching, concurrency control, naming, persistence, query, relationship, security, trading and transactions management services. Interfaces are provided to create objects, to control access to objects, to keep track of relocated objects, and to control relationships between classes of objects. These components provide the generic environment into which individual objects can perform their tasks. The purpose of the definition of the object services is to augment the bare communication facilities provided by the ORB with facilities that are generic in nature and can be used by more specialized components. So, facilities that OMG thought would be needed or would be useful in a distributed system have been specified. While nothing prevents a programmer from bypassing these facilities or using his or her own, their existence in a given context, and the fact that they have standard interfaces defined in IDL, will allow a developer to count on them if he wishes to, and use them instead of implementing everything from scratch. These facilities also offer a standard way for remote, unrelated objects written by different programmers to collaborate on some very elementary tasks. This elementary collaboration can take place indirectly through the facilities of public standardized services that are accessible from all objects. Most commercial ORBs, however, include an implementation of the naming or the trading service only.

The Common Facilities (OMG Facilities, 1998) are a set of components characterized by IDL interfaces, which provide common higher level functionality. These facilities are subdivided into two categories: the horizontal and the vertical Common Facilities. The horizontal facilities regard the user interface, the information management, the system management and the task management services. The vertical facilities, in contrast, are domain specific. For instance, the *Mobile Agent System Interoperability Facility* (MASIF) described in Chapter 7 is an example of a vertical common facility defined by OMG member companies specialized in agent technology.

Finally, the Application Objects are specific to end-user applications, and reflect user needs to define and develop distributed components for applications.

A distinctive feature of the OMA architecture is that the same object can alternatively play the client and the server roles at different times.

### 4.2.1.2    Common Object Request Broker Architecture

The CORBA specification (OMG CORBA2.3, 1998) provides a description of the interfaces and services that an OMA compliant Object Request Broker must implement in order to be aligned with the OMG standards. In addition, it defines a software infrastructure to facilitate the development of reusable and portable applications in a distributed environment.

The architecture, depicted in Figure 4.2, consists of various components, each responsible for a specialized functionality: the ORB, the IDL Stub and Skeleton, the

*Dynamic Invocation Interface* (DII), the *Dynamic Skeleton Interface* (DSI), the *Interface Repository* (IR) and the *Object Adapter* (OA).

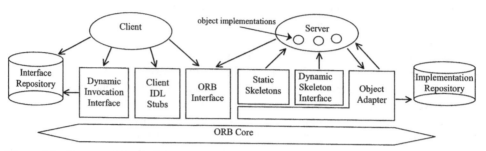

**Figure 4.2**  Common Object Request Broker Architecture

With the publication of the first CORBA 1.0 specification a number of commercial ORB implementations appeared. The fact that these implementations could not interwork emphasized the problem of ORB interoperability when objects residing on different ORBs needed to communicate. As was mentioned, this problem was overcome with the following CORBA 2.0 specification, that dealt explicitly with the interoperability issues by defining the Interoperable Object Reference format and the mandatory *Internet Inter-ORB Protocol* (IIOP), which specifies how *General Inter-ORB Protocol* (GIOP) messages have to be transported over a TCP/IP network. The elements of the CORBA 2.0 Inter-ORB architecture are the GIOP, IIOP and *Environment Specific Inter-ORB Protocols* (ESIOPs). GIOP defines a set of message formats and data type representations, which are unambiguously interpreted by all platforms, while IIOP specifies how GIOP messages are transported over the Internet. Although GIOP messages can also be mapped into other connection oriented transport protocols, an ORB must always support IIOP to be CORBA 2.0 compliant. Specific ESIOPs have also been defined for handling interoperability with non-CORBA compliant distributed platforms.

**Object Request Broker**
The ORB is the middleware responsible for establishing communication relationships between client and server objects in a distributed environment. It is in charge of identifying the server object that must satisfy the client request, activating it if necessary, passing input parameters and operation type, and returning the result and/or the output parameters. Each ORB implements a mechanism similar to remote procedure calls applied to objects. The CORBA specification requires that the ORB mediates all aspects of the communication between distributed objects, since the client must be aware neither of the server's location nor of the server's programming language. Clients need only care about an object's interface, as that is defined in CORBA IDL.

Before invoking any operation on a target object, the client must obtain its object reference, for example by interacting with the naming service or with other well known objects that are implemented to hold and return, when queried, the object reference needed. The ORB creates a new object reference once the target object has been activated, but the object itself is responsible for letting other objects find its reference by registering itself in a publicly accessible naming service, or by using any other custom distribution mechanism a programmer may choose to implement. The object reference can even be exchanged between two interacting objects in a string format contained in an e-mail or a file. Since the string format seems to be a very appropriate method of storing and/or passing object references around, the CORBA standard has foreseen the presence of two operations: the `object_to_string` and the `string_to_object`. The former allows an object reference to be converted into a string representation, while the latter does the opposite job. Inside a single vendor's ORB domain the object reference can have any proprietary format. However, whenever an object provides its services outside its local ORB domain, its object reference must have the standard IOR format specified in CORBA 2.0. Once an object reference is exported in a string format or published in a naming or trading service, it should be able to be stored and used after an arbitrary period of time, since the ORB guarantees that the associated target object will be available to satisfy client requests as long as it is active. Practically, depending on the product and the activation strategy employed, an object reference may become invalid and refer to no active object. In summary, an ORB offers to client objects the following services:

- *Object localization.* The client does not need to know if the target object is active within a process running on a remote machine, or in a process within the same machine, or even within the same process with the client. The ORB alone is responsible for localizing the target object using the object reference provided by the client. While ORBs are able to interpret object references according to the format they use, clients do not interpret IORs. They simply use them as opaque pointers in a flat communications space.
- *Object implementation transparency.* The client object need not be aware of the programming language used to implement the target object, or of the operating system or hardware platform on which the server program is executed. If fact, it has no way of discerning between implementations in different languages or on different platforms even if it wanted to.
- *Object execution state transparency.* The client does not need to know whether the target object is already active within the corresponding server process and ready to accept incoming requests. The ORB is responsible for transparently activating the target object before forwarding the client request.
- *Object communication mechanism.* The client can ignore the communication mechanism used by the ORB to send the incoming request to the target object or to return the response back to the client.

Thanks to these ORB capabilities, the programmer can concentrate on the problem domain of the specific application he needs to implement without worrying about low level networking issues.

**Interface Definition Language**
A client can determine the services offered by an object on the basis of its IDL interface. In fact, an interface describes the attributes and operations an object makes available to other objects, that is, a sort of contract between client and server objects. Interfaces in CORBA are defined in IDL which is a declarative language, with a syntax similar to that of C++, used to define *modules, interfaces, attributes, operations* and *data types.* It cannot be used to define any actual implementation, however, and is used as a programming language-neutral way to define method signatures and associated data types. There is a basic set of core data types, but the programmer is then able to define his own, composite data types

using the elemental ones. Mappings are defined between these data types and the actual data types supported by a real programming language. The OMG is responsible for specifying these mappings. So far mappings have been specified for most mainstream languages like C++, Java, COBOL, and so on. Methods defined in IDL are grouped into interfaces which can then be inherited. Again, the language mappings (also called bindings) describe how these concepts are expressed using the syntactic forms of an actual language. So bindings not only contain a mapping of data types, but also a mapping of concepts that appear in IDL into concepts of real programming languages. For instance, IDL defines a union construct which is easily mapped to a C++ union, but requires a more complex structure in Java which does not support unions at the language level. The same applies to most object oriented features of IDL like inheritance of interfaces. For modern, object oriented languages, mapping these features to concrete syntactic forms is a straightforward procedure. Since CORBA is also used with older languages like C or COBOL, this is not always the case.

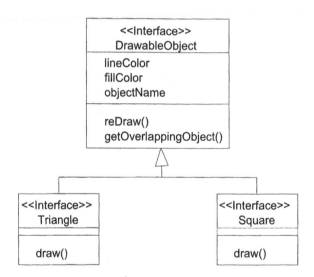

**Figure 4.3**  Triangle and Square interfaces inherit from the DrawableObject interface

A module defines a namespace where interface names must be unique. An interface defines a set of object attributes and operations. Attributes are object values which can be directly accessed for reading and writing or only for reading (read-only), using the standard get and set operations. Operations are methods whose signature is described in terms of an operation's name, its parameter modes (in, inout, out), names and types of its parameters, the type of its result and exceptions it can potentially throw. Finally, data types describe the types of values used for parameters, attributes, exceptions and return values. They can be primitive (boolean, char, double, float, long, octet, short, unsigned long, unsigned short) or structured types (any, enum, sequence, string, struct, union). In addition, the typedef construct allows the programmers to define application specific data types.

One of the key characteristics of IDL is that, being an implementation neutral language, it is not concerned with an object's implementation, but focuses only on the public methods and attributes of an object. After all, from a potential client's point of view, only these are relevant for an interaction. This is equivalent to saying that in a distributed environment clients should not be interested in implementation issues but only on offered services. During the specification phase, IDL files also provide useful material for documenting the software development process.

An IDL code example describing the interfaces of Figure 4.3 is given below. This example demonstrates the use of inheritance between IDL interfaces. It is shown that it is possible to define only once, in an abstract base interface, all common attributes and operations which derived interfaces can then simply inherit without re-declaring. The `reDraw` and `getOverlappingObject` operations are polymorphic, since they deal with generic objects that can be drawn (i.e. `DrawableObject`), while the `draw` operation's definition and implementation change depending on the specific object (i.e. `Triangle` or `Square`) to be drawn. This is unavoidable, since in the example given above, the argument list changes and so polymorphism cannot be employed.

```
module MyExample {
     interface DrawableObject {
          enum Color {red, green, blue};
          typedef sequence<string> StringList;
          exception OutOfDrawingArea {string reason;};
          exception ObjectTooBig {string reason;};
          exception NotExisting {string reason;};

          attribute Color lineColor;
          attribute Color fillColor;
          attribute string objectName;

          void reDraw (
               in string objectName);

          boolean isDisplayed (
               in string objectName
               )
               raises (NotExisting);

          void getOverlappingObject (
               in string objectName,
               out StringList objectNames
               )
               raises (NotExisting);
     };

     interface Triangle : DrawableObject {
          void draw (
               in unsigned short x_coord,
```

```
                in unsigned short y_coord,
                in unsigned short width,
                in unsigned short height
                )
                raises (
                        OutOfDrawingArea,
                        ObjectTooBig
                );
        };


        interface Square : DrawableObject {
                void draw (
                        in unsigned short x_coord,
                        in unsigned short y_coord,
                        in unsigned short lenght
                        )
                raises (
                        OutOfDrawingArea,
                        ObjectTooBig
                        );
                };
        };
```

**Static Invocation Interface**

Once the IDL definition of an object's interface is available, it can be used to obtain the implementation *skeleton* of the respective CORBA object, as well as code to be incorporated by client objects wishing to access it (*stub*). Naturally, the IDL specification is mapped onto a target implementation language using an IDL compiler. Once applied, this compiler will generate the necessary stub and skeleton files required for communication between the server and its clients. Stub and skeleton files are produced by the IDL compiler as source (or header where applicable) files for the target implementation language.

The skeleton files contain that portion of the server side's code, which implements the communication between the ORB and the server object. Any time the skeleton will receive a service invocation from the ORB, it will forward it to the object implementing the corresponding operation. To fully realize the server object, the programmer must write the actual code that implements the interface operations, i.e. the methods present in the skeletons.

The stub files contain that portion of a client's code which implements the communication between the client object and the ORB. When the client invokes an operation on a remote object, the stub is responsible for forwarding the service invocations to the ORB. The stub is also called a *proxy* since it represents the server object within the client execution environment (i.e. within the client process).

Figure 4.4 shows the relationship between the files produced by the IDL compiler and the client and server objects.

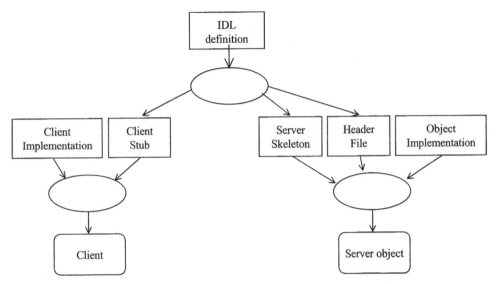

**Figure 4.4**  From the IDL definition to the client and server object implementation

Stubs and skeletons interact with the ORB for the marshaling and unmarshaling of the operation parameters in order to map them from the local programming language (e.g. Java, C++) format to a common format (i.e. the GIOP format used to guarantee ORB interoperability). The marshalling process consists of the coding of language dependent and data types into language and system independent data types, while the unmarshaling process corresponds to exactly the opposite procedure. The important thing to realize is that the IDL compiler is capable of producing stub and skeleton code once it is given the IDL files, because the only kind of information required for the stub and skeleton implementation is the operation signatures. Operation signatures (i.e. an operation's parameter types, the type of its return value and the type of exceptions it might throw) are all available in the IDL definitions.

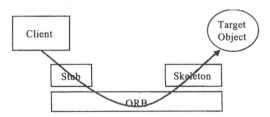

**Figure 4.5**  Static invocation interface

When stubs or skeletons mediate the interactions between the objects and the ORB, this is referred to as a *Static Invocation Interface* (SII) (see Figure 4.5). The term 'static' indicates that stubs and skeletons are respectively part of the client application and of the server object's implementation, and therefore the server object IDL interface must be known at client and server compilation time, as shown in Figure 4.4.

**Dynamic Invocation Interface and Dynamic Skeleton Interface**

In addition to the Static Invocation Interface based on stubs and skeletons, CORBA provides two more mechanisms: DII and DSI, for allowing dynamic method invocation (see Figure 4.6). In particular, with the DII mechanism a client does not at compile time need the target object's stub. It may be the case that at compile time the server object's IDL definition is not available and becomes known only at runtime. Generally, DII removes the necessity to have an *a priori* knowledge of the server object's interface. Instead, it is possible to have the signature of an operation, including its name and the type of its parameters, become known only at runtime. At that point, the method invocation can be dynamically constructed. The DII mechanism exploits a set of ORB interfaces, which are independent from the server object's IDL interfaces. To support these features CORBA has introduced standard APIs to be used with DII. Quite symmetrically, the DSI mechanism allows the server object to be unaware at compile time of the object skeletons, i.e. of the object IDL interface, and therefore of the specific type of object it is implementing. Again, CORBA has defined interfaces for that purpose providing a runtime binding mechanism for DSI. Instead of invoking a specific method using a statically compiled skeleton, the same generic method is used all the time. This generic method is called *Dynamic Invocation Routine* (DIR), and can accommodate many different interfaces. The server object in the DSI case extends no specific IDL compiled skeleton but a generic base class provided by CORBA (`DynamicImplementation`). The same class is used by all dynamic implementations irrespectively of the server object they implement. This is in contrast with the static skeleton approach, where each server object extends a tailor-made class (the skeleton class) specifically produced with a certain IDL interface in mind. With the static skeleton approach, the interface implemented by a server object is determined at compile time, where the skeleton class to be extended (and hence the interface) is fixed. In the dynamic skeleton approach it is up to a server object to implement any type, and the interface it actually supports is determined by the runtime behavior of the object. Similarly, with the static skeleton approach, type checking is more strict and is provided by the ORB itself. In the dynamic skeleton approach, type checking has to be implemented by the client programmer. So, for instance, if when using an SSI, a client holding an object reference uses it to invoke a method that does not exist, the ORB will throw a `BAD_OPERATION` exception[1]. When using DSI, on the other hand, it will be the responsibility of the programmer to throw that exception if he or she determines that a wrong method was invoked. The ORB is not able to determine this itself, since it perceives no type information but a generic interface (DIR) to be invoked at all times. In essence, the price to be paid by the DSI mechanism for its flexibility is that it is, by definition, weakly typed. Instead of invoking specialized methods directly and in a manner that would allow for compiler checks, a generic interface is used in all cases, and the specific info (e.g. the name of the method) is passed as a string. This forfeits the advantage of compiler type checking, but also circumvents restrictive compiler rules. The same trade-off applies when comparing the dynamic invocation interface with a stub solution. It is useful to note here that the dynamic skeleton and dynamic invocation interfaces are completely orthogonal to each other. In fact, all four combinations make sense and can be used (DII with DSI, SII with DSI, DII with DSI and SII with SSI). More importantly, in the same way in which a server cannot tell whether a certain invocation came using a dynamic or a static invocation interface, a client

---

[1] Under normal conditions this error would have been reported at compile time.

cannot tell whether the server object it is interacting with is implemented using a statically compiled skeleton or the dynamic skeleton approach. A further point to be made is that a scenario using DII with DSI represents a low level solution that very much resembles an equivalent message-based protocol one (e.g. where a custom application protocol operates on top of sockets).

In conclusion, the added flexibility that DII and DSI provide makes it is possible to accommodate new dynamic services where the operations to be invoked and their parameters are not known until runtime. Examples include general interoperability solutions and interactive software development tools that call for a high degree of flexibility and dynamic behavior.

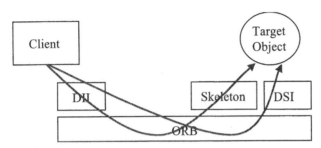

**Figure 4.6**  Dynamic invocation interface and dynamic skeleton interface

**Interface Repository**
Once a client has obtained a reference to a remote object it usually downcasts this reference to allow specific methods to be invoked. This presupposes that knowledge of the object's specific interface is available at compile time. Assuming that the client has only at runtime discovered the object and has no *a priori* knowledge of its interface, the alternative is to invoke the object's `get_interface` method. That method is part of the definition of the interface `Object` which is implemented by all CORBA objects, so is available without requiring any downcasting. What the method actually returns is the reference to another remote object that contains information about the IDL definition of the original object. The object holding this meta-data information implements the `InterfaceDef` (from interface definition) interface, and is stored persistently using the facilities of an Interface Repository.

So, the Interface Repository is a CORBA object that manages a persistent, runtime database containing the IDL definitions of objects residing in the ORB. The Interface Repository can be invaluable when using the DII mechanism. Recall that, according to that technical information about the server, an object's IDL interface is not fixed in the stub code to be used, but is instead available only at runtime. The client application is thus coded in such a way as to use this information meaningfully and obtain the necessary methods. Essentially, the interface of the server object is only manifested by the runtime behavior of the application code. Since the Interface Repository stores persistent objects holding this IDL information, active objects query it to get information about registered interfaces. In this way, it can offer a more structured way for an application to obtain information about server objects, learn their interfaces, the names of their methods and the types of their parameters and then manage to invoke methods on them. With CORBA 2.0, support for federated Interface Repositories and for runtime updating has been added. Since the Interface Repository can contain a huge amount of self-describing meta-data information it can serve several other purposes. For instance, it can be used by the ORB to check at runtime method signature correctness, to provide meta-data information to clients, or to be used with interface browsers or other debugging tools.

**Object Adapter**

An object implementation takes advantage of ORB services through the *Object Adapter* (OA), which can be thought of as the glue that connects object implementations to the software bus. In particular, the Object Adapter provides a set of necessary, low level services like object implementation activation and deactivation, server class registration in the *Implementation Repository*, object reference generation, incoming service invocation forwarding to the interface skeleton, and incoming request dispatching to the target object. Since these facilities may require different interfaces depending on the environment, it would not be feasible to have the ORB provide a single interface useful to all. So, in effect the ORB interface contains only those methods that are expected to be context-independent, and all functionality that may be subject to change depending on the specific circumstances is provided by means of object adapters, tailored to specific needs. In that respect, one could imagine object adapters as an extension mechanism for the ORB core. The Implementation Repository has a purpose analogous to that of the Interface Repository. While the latter is mostly useful for client objects wishing to discover a server object's interfaces, the former is paired with the Object Adapter. It contains implementation activation information about objects the Object Adapter may need to activate. It is a runtime database containing the object references, types and classes of all objects, which have been instantiated by each server process. The ORB uses this information to locate active objects and to request the activation of new objects within a server. Since an OA defines how an object must be activated, different OAs could be implemented to support the different object activation policies (e.g. new thread, new process, existing thread) that may be required by an application. To avoid a proliferation of OA implementations, CORBA has defined a *Basic OA* (BOA) which encompasses all of the usually needed features and should be appropriate for most cases. The BOA implements the following activation policies:

- *Shared server.* According to this policy, the server is activated the first time an operation is invoked on one of its objects. Once the server has finished initializing, it informs the ORB that it is ready for processing incoming calls by invoking the connect operation. From that time on, the same process will manage all incoming invocations irrespective of the object to which they are addressed.
- *Unshared server.* Here, a new server process is spawned for each incoming call to an object instance managed by the server. Similar to the previous policy, once an object has completed initialization it notifies the ORB that it is ready. This policy can result in object instances implementing the same interface being executed within different processes.
- *Server per method.* This policy pushes the unshared server strategy to the limits; here, a new server process is activated for each method invocation. In this policy, the ORB does not need to know if an object is already active at the time when a request for it arrives.

- *Persistent server.* The server here is not activated by the BOA, but is typically activated manually (the server process has to invoke the connect operation). In most other respects it is similar to the shared server activation mode.

The *Portable OA* (POA) has been introduced in CORBA 2.2 with the intention that it should handle the problem of object implementation portability across different ORB products. In addition, POA, which is intended to substitute for BOA, supports the following features (OMG CORBA 2.3, 1998):

- Increased support for persistent objects. These objects are responsible for providing consistent services to clients holding references to them even though their lifetimes may span different server activations by the object adapter.
- Support of both explicit and on demand activation modes.
- Transparent activation of objects (this is also related to the previous points).
- Support for transient objects.
- Free the ORB from the need to maintain any particular persistent information describing individual objects.
- Introduce the concept of a servant, that provides realization for a CORBA object and describes an entity called a 'servant manager' that co-operates with the ORB to manage association between objects and servants.
- Allow for multiple instances of a POA to exist in a server and provide a mechanism (Adapter Activator) to activate additional POAs on demand.

### 4.2.1.3   CORBA Services
The updated list of CORBA services described in OMG Services (1998) is as follows:

- *Naming Service.* Provides a mechanism for locating other objects on the basis of a target object name. Human readable names are assigned to objects and are used to uniquely identify objects within a naming context and to find their object references (*Interoperable Object Reference* – IOR). In this sense the naming service implements a white pages functionality. The naming space is not flat but is defined in a tree-like structure, as naming contexts can be nested within other naming contexts to arbitrary depths. The fully qualified name of an object has to be read from the root context down to the individual leaf representing that name in a Domain Name System fashion. Since the Naming Service is defined as a CORBA object with a published IDL, all objects residing on an ORB can access its services provided one exists. Moreover, due to the large amount of information that a Naming Service may hold, federation of Naming Services is allowed by allowing a naming context to be associated with an object reference on a different machine.
- *Event Service.* Allows objects to dynamically register and de-register their interest in specific events. An object (consumer) which is interested in being notified upon the occurrence of a certain event does not need to explicitly request this notification from the object (supplier) that produces it. Instead, it instructs the event service as to which events it is interested in receiving and delegates any further responsibility to it. So the role of the event service is to mediate, through an event channel object, the notification delivery process in order to leave suppliers and consumers independent from each other

and to support asynchronous communications. Consumers and suppliers are interacting with the event service and the latter brokers between the two. Two communication models have been defined: the push and the pull models. The first foresees that the supplier pushes the event to the event channel, which in turn pushes it to the registered consumers. While according to the latter, the consumer pulls the event from the event channel, which in turn pulls the event from the registered suppliers.

- *Lifecycle Service.* Defines conventions for creating, deleting, copying and moving objects. Since CORBA-based environments support distributed objects, lifecycle services define services and conventions that allow clients to perform lifecycle operations on objects in different locations.

- *Persistent Object Service.* Provides a set of common interfaces to the mechanisms used for accessing and managing the persistent state of objects. While a CORBA object is ultimately responsible for managing its state, it can use the Persistent Object Service or delegate to it some of the responsibility for the actual work.

- *Transactional Service.* Provides a mechanism for treating a set of object interactions as a unique transaction. The transaction model describes when a transaction starts, when it ends and how it can be recovered. In this way, an entire series of operations can be viewed as an atomic operation that is either brought to the end, or if a problem arises while it is in an incomplete state, the effects of individual operations that have taken place so far can be recovered and the state of the system fully restored. The transaction service supports flat and nested transactions even across different ORBs. The objects involved in a transaction can assume the following roles: transactional client (which begins and ends a transaction); transactional server (which is affected by the transaction but is not interested or cannot handle a recovery); and recoverable server (which manages data and is therefore affected by the transaction commitment or roll back). The transactional service implements the standard two-phase commit model.

- *Concurrency Control Service.* Enables multiple clients to coordinate their access to shared resources. Coordinating access to a resource means that when multiple, concurrent clients access a single resource, any conflicting actions by the clients are reconciled so that the resource remains in a consistent state.

- *Relationship Service.* Allows entities and relationships to be explicitly represented. Entities are represented as CORBA objects. This service defines two new kinds of objects: relationships and roles. A role represents a CORBA object in a relationship. The Relationship interface can be extended to add relationship specific attributes and operations. In addition, relationships of arbitrary degrees can be defined. Similarly, the Role interface can be extended to add role specific attributes and operations.

- *Externalization Service.* Defines protocols and conventions for externalizing and internalizing objects. Externalizing (also called serializing) an object is to record the object's state as a stream of bytes (in memory, on a disk file, across the network, and so on). Internalizing (also called de-serializing) an object is to use the externalized representation in order to recreate the object back in the memory space of a different (or the same) process. The externalized object can exist for arbitrary amounts of time, be transported by means outside of the ORB, and be internalized in a different, disconnected ORB. For portability, clients can request that externalized data are stored in a file whose format is defined by the Externalization Service Specification. This allows an application to internalize an object even though it didn't have control over its externalization (since an agreed-upon format was used).

- *Query Service.* Allows users and objects to invoke queries on collections of other objects. The queries are declarative statements with predicates, and include the ability to specify values of attributes, to invoke arbitrary operations and to invoke other Object Services.                                                                    .
- *Licensing Service.* Provides a mechanism for producers to control the use of their intellectual property. Producers can implement the Licensing Service according to their own needs, and the needs of their customers, because the Licensing Service is not imposing it own business policies or practices.
- *Property Service.* Provides the ability to dynamically associate named values with objects outside the static IDL-type system. It defines operations to create and manipulate sets of name-value pairs or name-value-mode tuples. The names are simple OMG IDL strings. The values are objects weakly typed after the OMG IDL 'any'. The use of type any is significant in that it allows a property service implementation to deal with any value that can be represented in the OMG IDL type system. The modes are similar to those defined in the Interface Repository's AttributeDef interface.
- *Time Service.* Enables the user to obtain the current time together with an error estimate associated with it. It can also ascertain the order in which 'events' occurred, generate time-based events based on timers and alarms, and compute the interval between two events. It consists of two services, the Time Service and the Timer Event Service.
- *Security Service.* Handles identification and authentication of principals (human users and objects which need to operate under their own rights), authorization and access control and security auditing to make users accountable for their security related actions.
- *Object Trader Service.* Allows client objects to locate server objects on the basis of the requested service (i.e. on the basis of the requested IDL interface). The trader manages the associations between the descriptions of the type of an offered service and its object reference. This description specifies the operations supported by the object, the types of their parameters and return values. It also manages some additional information, which is useful to determine the target object. The trading service can be thought of as the object yellow pages, and in this respect complements the Naming Service (white pages).
- *Object Collection Service.* Provides a uniform way to create and manipulate the most common collections generically. Examples of collections are sets, queues, stacks, lists and trees. For example, sets might support the following operations: insert new element, membership test, union, intersection, cardinality, equality test and emptiness test.

### 4.2.1.4    Standardization

In the last few years, the Distributed Object Technology has been adopted as one of the key technologies to support the development of distributed applications. In this context, CORBA has been established as an important standard, enhancing the original Remote Procedure Call based architectures by allowing a relatively straightforward and transparent realization of distributed service functionality. The increasing number of success stories, that are related to the use of this standard in different industries, and in very different ways, have propelled CORBA to its present position as the *de facto* standard in building enterprise-wide computing systems. OMG continues to build on these successes by making improvements to the standard. This has led to the definition of the CORBA 3.0 specification. This section presents a short history of the CORBA standardization efforts, and provides an overview of the features supported by the new version.

## History of Specifications

The CORBA specifications have undergone several modification steps. At this moment, OMG members are working on the CORBA 3.0 version. The following list summarizes revisions made to CORBA standards over the past years:

- *CORBA 1.0 (October 1991).* It included the Common Object Request Broker Architecture Object model, Interface Definition Language, and a first set of Application Programming Interfaces (APIs) for the dynamic request management and invocation (DII) and the Interface Repository that provided persistent storage of IDL interfaces, modules, and other IDL type definitions. It also defined the mapping of IDL constructs to the C programming language.
- *CORBA 1.1 (February 1992).* This version was the first of the Common Object Request Broker Architecture specifications to circulate widely. The contents of the specification differed with respect to those of the previous one in that many ambiguities were removed, new interfaces for the Basic Object Adapter and memory management were added, and that certain issues concerning the object model and the Interface Repository were clarified.
- *CORBA 1.2 (December 1993).* This was mostly an update version, removing several ambiguities, in topics such as memory management and object reference comparison.
- *CORBA 2.0 (August 1996).* This version represented a major improvement to the standard. While preserving the existing CORBA object model, the new version introduced several new, major features. Included in this release were an Interoperability Protocol specification, Interface Repository enhancements, ORB initialization, and the IDL bindings for the C++ and Smalltalk programming languages. Other improvements and extensions comprised:
  - The Dynamic Skeleton Interface mechanism
  - Extensions to the Interface Repository
  - Specification of an initial reference resolver to ensure client portability
  - An interoperability architecture with the definition of GIOP, IIOP, and DCE CIOP
  - Inter-working with OLE2/COM
  - Support for layered security and transaction services
  - Data-type extensions for COBOL, scientific processing, and wide characters.
- *CORBA 2.1 (August 1997).* This version enhanced the previous version by adding additional security features such as secure IIOP and IIOP over SSL and language bindings for COBOL and ADA. Modifications and revisions of the interoperability aspects were also included, as well as extensions to IDL types.
- *CORBA 2.2 (February 1998).* CORBA 2.2 included the Server Portability enhancements and the definition of the Portable Object Adapter. The most tangible offering of POA was that it allowed CORBA servers to run on a different CORBA platform without requiring re-compilation. Additional enhancements were the specification of an inter-working scheme with DCOM and the definition of the IDL to Java language mappings.
- *CORBA 2.3 (September 1998).* This is the currently effective version, which was basically a maintenance release over 2.2. It comprised revisions to the IDL/Java language mapping, ORB portability, COM/CORBA Part A and B, CORBA core, ORB Interoperability, and CORBA Security 1.2.

**Latest Version**
The CORBA 3.0 version was officially announced at the Comdex Enterprise '98 Object World Conference. This version offers a new range of opportunities for both seasoned distributed computing programmers and less technical business programmers. It simplifies the use of CORBA ORBs for the development of distributed object applications. It is already seen by many vendors as a landmark specification, into a new era of Distributed Object Technologies. Newly added CORBA 3.0 capabilities fall into the categories below:

- Support for distributed components
- Quality of Service features and new Messaging support
- Other technologies that enable complete Internet integration and support for legacy environments.

A pre-production release of this version was made available to CORBA developers after the OMG meeting in San Jose, California, on August 27, 1999, giving them the opportunity to exploit the new features or implement new CORBA services prior to the final revised release of CORBA 3.0, scheduled for mid-2000.

*Distributed components*
One of the most significant additions to CORBA was the specification and support of components. Generally, a component is a reusable software object that is part of a framework for developing applications, following the plug-and-play philosophy. A component can contain other components within. This composition of instances is dynamic, and any relationship among the components can change at runtime. Components can be assembled into a larger component and ultimately into an application. For usage purposes, the composed component or the application must be instantiated and initialized. The component specification is subdivided into four major areas:

- The CORBA Component Model (OMG Component, 1999). It specifies interfaces and mechanisms for a CORBA-based component model and for the mapping of that model into other component models such as Sun's Enterprise Java Beans technology. This model provides a more complete mechanism for assembling objects into applications. Besides, by having such a model, programmers are able to build components that will be capable of inter-operating with other emerging component technologies, i.e. it enables a tighter integration with other components, making it easier for programmers to use CORBA. A CORBA component is a container of CORBA objects. Such a component can provide multiple interfaces, called facets, which are capable of supporting distinct CORBA interfaces. Since a CORBA component can be persistent and transactional, the specification defines a multi-platform software distribution format, using XML. Also, as it is described within the specification, mechanisms and interfaces are defined for serializing a component's state, and for constructing a component from the serialized state. The serialization mechanism is suitable for storage and retrieval, and for externalizing state over a communication channel, i.e. allowing object migration between clients and servers without loss of state.
- The CORBA Scripting Language (OMG Scripting, 1998). This specification makes the composition of CORBA components to applications easier by using scripts. A CORBA script minimizes the complexities of an application's development process, for example,

by removing the need for memory allocation and compilation. Hence, this will speed up programming, especially of a prototypical version of an application.

- The CORBA Multiple Interfaces. In contrast to previous CORBA specifications, where a CORBA object can only realize a single interface, a CORBA component allows the composition of multiple interfaces. In this way, a means for objects to be composed of logically distinct services by the use of multiple interface definitions is provided. A program can also control access to the functions of an object based on interface definitions. This specification is part of the OMG Component (1999).
- The CORBA Object-by-Value (OMG Object, 1998). Object by value is a straightforward and efficient method for passing CORBA objects as parameters in CORBA object operations. Pure CORBA 2.2 objects cannot be passed by value; only pass by reference semantics are supported. However, there are many occasions in which it is desirable to be able to pass an object by value, rather than by reference, allowing an application to explicitly make a copy of an object. By applying the pass by value feature as described in Objects-by-Value, the receiving side of a parameter passed by value would receive a new instance of an object that has no relationship with the object it originated from, at the sending side. This feature allows a designer to specify location-aware applications, in which composed objects are able to move around between client and server. The benefits of this mechanism can include speeding up the interaction between a client and a server. It must be clarified here that user-defined data structures used as parameters in method operations can already be passed by value. A CORBA object, however (a remote accessible object implementing an interface), cannot.

*Quality of service*

The quality of service part of CORBA 3.0 provides the means to adapt CORBA to various underlying transport systems and/or operating systems, depending on the distributed application's needs and the resources provided by the environment. The aim of this specification is to extend the applicability of CORBA, and its ease of use. The Quality of Service specification comprises three major extensions to the previous CORBA version. It includes:

- CORBA Minimum ORB (OMG Minimum, 1998). It describes a CORBA environment for systems with limited resources. The specification defines a single profile, a subset of the CORBA reference model cut down from CORBA 2.2. Its intention is to meet the needs of a CORBA compliant system that operates in an embedded system while preserving the key benefits of CORBA: portability of applications and interoperability between ORBs. Such a system naturally requires more sophisticated management decisions on the use of its (scarcer) resources, and tends towards a much more predictable system environment. As a result the approach adopted by this specification is to identify a scalable process which allows the configuration of the CORBA ORB down to a core minimum set. The proposed Minimum CORBA fully supports the IDL. Certain core components of other specifications, however, like the Interface Repository, the Dynamic Invocation Interface, the Dynamic Skeleton Interface, and interoperability aspects with DCOM and DCE, were left out of the specification.
- CORBA Real-Time (OMG Real-Time, 1998). It describes an ORB system to be used in a real-time configuration. This specified architecture describes an extension to CORBA called the Real-time ORB. To meet the real-time requirements, the specification

provides a framework that allows developers to design applications which meet the timing requirements set and the criterion of end-to-end predictability of activities in the system. Key to this framework is providing to the programmers access to the management of the system's resources. To this end, the specification exposes interfaces that allow developers to control resource allocations within the ORB. These interfaces are not available in more typical configurations.

- CORBA Asynchronous Messaging and QoS Control (OMG Asynchronous, 1998). CORBA in its present form provides for synchronous, deferred synchronous and one-way messaging. With the extension of asynchronous messaging, CORBA is providing a mechanism for managing asynchronous messages in a distributed system. Asynchronous messaging allows a client to invoke a method on an object without having to block for the return. The invocation behavior can be constrained or parameterized by setting certain service parameters that govern the messaging quality of service aspects. At this moment, the available parameters include:
  - Delivery Quality. Specifies the characteristics of the delivery of a message, e.g. based on expire times for request and reply or reliability aspects.
  - Queue Management and Message Priority. Requests at the server side can be queried or ordered based on temporal or priority indications.

*Internet integration and legacy support*
The primary goal of CORBA 3.0 is to simplify the use of CORBA technology for the development of distributed object applications and extend the business scope of CORBA. In the last few years, the Internet networking community has become one of the fastest growing in all of the telecommunications information sectors. OMG has recognized this and has responded with the provision of a CORBA version with support for a thorough Internet integration. Also, to increase the acceptance of CORBA by other technology providers, support for legacy environments was also enhanced:

- CORBA Java to IDL Mapping (OMG Java, 1998). The Java language to IDL mapping allows Java application developers to build distributed applications purely in Java and then generate the appropriate CORBA IDL from compiled Java class files. This process of 'IDL de-compilation' allows easy integration with CORBA without the need to write IDLs, and allows other CORBA-compliant applications access to Java applications over the Internet Inter-ORB Protocol (IIOP).
- CORBA Firewall (OMG Firewall, 1998). Many applications are behind a firewall for security reasons. This specification defines a set of interfaces for allowing IIOP requests and replies to pass through a firewall. It encompasses a configuration mechanism for instructing the firewall to perform filtering and proxying on the client and server side. This is a very important feature in that it enables controlled and secure use of CORBA-based applications on the Internet.
- DCE/CORBA Inter-working. With the intention of increasing CORBA's acceptance, and in order to facilitate the integration of DCE legacy equipment into the CORBA bus, the DCE/CORBA Inter-working specification was defined. It provides specifications for an application level inter-working between CORBA clients interacting with DCE servers under different scenarios. The opposite case (DCE clients contacting CORBA servers) is also described. A unification of CORBA services implemented using DCE services is also foreseen in the specification. The business benefits for organizations with

investments in DCE technology are obvious. They can dock DCE applications into the CORBA world and ensure a prolonged use of their DCE legacy applications.
- CORBA Interoperable Naming Service (OMG Interoperable, 1998). Until recently, pure CORBA 2.2 clients could access a CORBA object's interface only through its CORBA reference. There was no other mechanism to contact a server, even if the client knew its location. The Interoperability Naming Service Specification defines a URL-like object reference format that can be used to reach known services at remote locations.

### 4.2.2   Microsoft architecture

Microsoft's Component Object Model represents the *Object Linking and Embedding* (OLE) object bus which is an integration scheme to support OLE2 components. It was originally designed to work in a single machine environment and only with the Windows operating system. In fact, it has been introduced to support the linking of Microsoft Windows 3.1 compound documents (OLE documents). Later, to allow distributed OLE components to interact with remote components executed in different machines, Microsoft produced Distributed COM. When referring to the Windows distributed components architecture, ActiveX has to be mentioned. ActiveX deals with the integration of components in Windows applications such as Internet Explorer (the Microsoft Web browser) and other applications.

DCOM provides a middleware for distributed computing based on the Microsoft version of *Distributed Computing Environment* (DCE) Remote Procedure Calls that represents an alternative to CORBA. In particular, DCOM and CORBA provide a solution to the same problem by exploiting two different object models, as will be explained in this section.

The COM architecture comprises a set of components including the object bus and the object services. The object bus is a distinct COM component while the object services, which were originally non-compound document elements of OLE, are not completely part of COM because some of them are merged with the Windows operating system. Typically, COM services were originally OLE Custom Controls like connectable objects, event sets, etc.

The COM object bus provides the same functionality as that of the CORBA ORB, but in a different way:

- It supports the concept of an interface but distinguishes between the object interface, which is defined using the proprietary Microsoft IDL, and the object implementation.
- It supports component reuse but via containment and aggregation rather than inheritance. In fact, Microsoft IDL does not support multiple inheritance between interfaces.
- COM objects do not maintain state information. They offer the services, which are available on the basis of the interfaces they support, and the client refers only to such functions and never to objects.
- It handles static and dynamic interfaces. COM provides a Type Library that is used to retrieve the interfaces, the operations and parameters supported by objects, and also some registry and object look up services which allow for the determination of an object's implementation.

Analyzing COM in more detail, it is possible to recognize that the concepts of interface, object and server are defined in a way that is quite different to CORBA (Orfali, 1996).

A COM interface specifies in a language independent way a client-server contract in terms of member functions and parameters. Unlike CORBA, no binding to specific languages has been specified. An interface is defined by a low level binary API, which operates on a table of pointers. Clients refer to server implementation methods through a pointer to the Virtual Table (which is simply an array of function pointers).

A COM object is typically a sub-component of an application, which represents a point of connection to other parts of the application or to other applications. It is an OLE object that supports one or more interfaces as foreseen by its class definition. The class implements the interfaces, and it is characterized by a globally unique identifier, the Class ID. A COM object is therefore a runtime instantiation of a class whose functions can be invoked by clients only through interface pointers. All COM objects must implement the IUnknown interface that is used by clients to control the object lifecycle, and to get pointers to the interfaces supported.

Finally, a COM server is typically a library (DLL) or an executable (EXE) file implementing object classes. As soon as a client requests an object of a given Class ID, COM runs the corresponding program and asks it to create a new object.

Microsoft has recently produced a new version of COM, called COM+, which provides new features like multiple interface inheritance and runtime object management, and which is accompanied by language oriented development tools.

## 4.2.3    Comparison of architectures

It is not an easy task to compare CORBA and DCOM, since these two architectures have many points in common but also many differences. Therefore, it is necessary to concentrate on a few aspects, which could help us first in the understanding, and later in the comparison and evaluation, of the two approaches. A number of technical reports and research papers (Tallman, 1998;Yee, 1999;Chung 1998) address this topic. Here, we concentrate on the following aspects:

- *Specification.* The approaches adopted by OMG and Microsoft are really different. The CORBA architecture is based on a well defined set of specifications. Although these specifications are produced by an industrial consortium and not by a public organization's standards committee, they nevertheless enjoy the status of *de jure* standards. On the other side, the DCOM architecture is the result of the efforts of a single (although market leading) private company which has produced a *de facto* standard for Windows environments. Unlike CORBA, DCOM is not fully specified, and therefore it is not always possible to completely understand the architectural characteristics starting from its public specifications. Moreover, the Microsoft implementation does not allow one to make a clear distinction between COM and Windows32 API functions. On the other side, even if CORBA specifications are complete and consistent, they are not always so detailed as to guarantee a compatible implementation by different vendors.
- *Interoperability.* COM has been designed for binary compatibility. In fact, it clearly separates interface definitions from their implementations in order to allow clients to

select at runtime the object interfaces to be used. The client-server interoperability is guaranteed at a binary level. On the other hand, CORBA allows interoperability between applications written in the major programming languages thanks to the specification of the mappings of IDL into different languages. That is, client-server interoperability in CORBA is guaranteed at a syntactic level (the IDL level). Nevertheless, both models draw a distinction between an interface and its implementation.

- *Language Support*. While CORBA is founded on the use of IDL for both system and user-defined interfaces, Microsoft considers IDL simply as one of the possible tools for defining them. In fact, OMG has defined not only IDL, which provides a generic framework for defining interfaces using both built-in and user-defined data types, but also its mapping into different programming languages like C, C++, Java, Cobol, etc. Microsoft instead has not defined a common set of data types accessible by all target languages. In fact, common data types called variant compliant have been defined only for OLE automation. The release of COM+ which provides a wizard tool-based workaround will probably allow users to overcome this problem.

- *Platform Support*. The COM solution is clearly targeted at the Microsoft environment, in spite of the fact that Microsoft at first encouraged other vendors to port it to different platforms and later, as this resulted in a partial failure, decided to take care of this activity itself. However, for the time being, it is not clear if this Microsoft commitment will persist along with the COM/COM+ product evolution. Regarding Java, it is clear that the effort done by Microsoft for extending its non-standard Java Virtual Machine (JVM) to provide intrinsic COM support is advantageous only for Microsoft Java developers. On the other hand, CORBA implementations are available for the most widespread operating systems and standard programming languages.

The evaluation of the pros and cons of the two solutions is left to the reader. However, it is reasonable to expect that in the near future, CORBA will continue to be the reference solution for large scale and multi-platform systems, while COM will dominate in the pure Windows environments.

It is important to note that since the release of CORBA 2.0, OMG has been faced with the problem of CORBA/COM interworking. This problem has been addressed by defining an interworking architecture able to allow for two-way communication between CORBA and COM objects. Objects residing in one object model are viewed as if they existed in the other object model. Since some CORBA products that implement this feature are available off the shelf, it is possible to realize a CORBA/COM mixed solution with moderate effort.

## 4.3    Distributed Object Technology in Telecommunications

The adoption of Distributed Object Technologies can lead to a more flexible and efficient telecommunication environment, capable of supporting advanced multimedia network services. In fact, the concept of openness in networks can be well served by distributed object architectures such as CORBA and DCOM. The term 'open network' reflects the exposure of key interfaces, which offer useful abstractions of network resources and generic network services.

DOT can be considered as the medium for providing these open interfaces with its neutral language specification (IDL) and reusable library of object services (CORBA services) that facilitate rapid deployment of network services. Moreover, the remote accessibility that these interfaces enjoy fits well with the concept of a network that provides access to much of its functionality or to resources of its internal nodes to its potential users. More specifically, the adoption of Distributed Object Technology can offer the following benefits to telecommunications systems (Redlich, 1998):

- *Abstraction.* The hierarchical structure of an object oriented information model offers different levels of abstraction to both the network services and user applications. So, a multimedia user application, for example, does not have to know the details of a QoS-sensitive service providing entity, or of the underlying communication infrastructure.
- *Location Transparency.* A distributed object architecture coupled with the use of a naming service facility can actually free application objects from the need to be aware of any physical locations, and makes the provision of programmable objects easier and completely transparent. This location-independent access to objects may be used, for example, in third-party communication session set-up, a feature that is difficult to realize with standard network signaling protocols.
- *Modularity.* It is a built-in feature of object technology. Object methods can be invoked by a number of different network services and applications designed separately from the particular object, which is a self-contained entity tightly integrated with the data it manages. In this way, an entire third-party or in-house developed hierarchy of services can be put to other uses and exploited rapidly.
- *Software Reusability.* Libraries of supporting objects can be provided together with a distributed object platform or as separate software. As long as IDL definitions of the services are available, integration with other pieces of software is seamless. As examples of this kind of software library, one can consider the defined CORBA Services and Facilities, which offer functions such as security, object lifecycle management and various special purpose facilities that otherwise should be rewritten for all applications using them, increasing in this way the time and the cost needed for application development.

Experience over the past few years demonstrate that Distributed Object Technologies such as CORBA are well suited to best-effort, client-server applications running over conventional local area networks. However, many application domains, such as telecommunications, require real-time guaranties from the underlying networks, operating systems and middleware components to meet their QoS requirements. Building real-time and QoS sensitive applications with CORBA is much harder than simple client-server applications, since the developer is given less access to lower layer protocols and is not able to implement customized procedures that may speed up an application in a particular context. The CORBA standard (as well as many vendor-specific implementations which usually include additional, non-standard features) does not address non-functional requirements which are nevertheless essential in many fields of distributed computing. Such requirements can include real-time ones, quality of service guarantees, high-speed performance and fault tolerance.

The absence, in CORBA specifications, of policies and mechanisms for providing end-to-end QoS guarantees is a serious impediment to the establishment of a CORBA-based telecommunication management and control infrastructure. For instance, there is currently no standard way of indicating relative priorities on objects' requests or of informing the ORB of how frequently an isochronous object service should be executed. Hopefully, many of these shortcomings will be overcome with the release of CORBA 3.0.

Another issue is the lack of built-in performance optimization mechanisms. Existing ORBs incur significant runtime throughput penalties and latency overheads. These overheads include excessive data copying, non-optimized presentation layer conversions and inefficient de-multiplexing algorithms.

Even though these issues are quite important and present legitimate sources of concern, more and more telecommunications industries are facing the prospect of replacing part of their existing telecommunications software infrastructure with novel, CORBA-based systems. This is not unrelated to the advantages in rapid service provisioning the adoption of DOT can offer.

Recently, OMG produced a first set of specifications, explicitly targeted for telecommunications. The CORBA Telecoms book (OMG Telecoms, 1998) is a domain technology document defining IDL interfaces pertinent to telecommunications systems. In the current version, specifications for the control and management of audio/video stream have been introduced.

What CORBA Telecoms basically does is propose a set of interfaces that would be useful in the implementation of a distributed, media-streaming framework. The principal components of this framework are:

- *Multimedia Devices and Virtual Multimedia Devices*. A multimedia device abstracts one or more items of multimedia hardware and acts as a factory for providing virtual multimedia devices. A multimedia device can support more than one stream simultaneously.
- *Streams*. A stream may contain multiple flows. Each flow carries data in one direction, therefore the endpoint may be either a source or a sink. An operation on a stream, such as start/stop, can be applied to all flows within the stream simultaneously, or to just a subset of them.
- *Stream Endpoints*. In relevance with the stream, a stream endpoint may contain multiple flow endpoints.
- *Flows, Flow Endpoints and Flow Devices*. There can be a CORBA object representing a flow connection and endpoint. Each flow endpoint can be included in a stream connection or it can be dangling. A flow device is exactly analogous to a multimedia device and is associated with a specific flow.

An example of a simple streaming architecture based on CORBA Telecoms is given in Figure 4.7. This figure presents a streaming architecture handling audio/video streams from one end-point (source) to another (sink). The end-points contain three basic entities:

- *Dataflow point,* which is the final destination of the data (stream) flow.
- *Control Object,* which is the Stream Interface Control Object responsible for controlling the data stream. Additionally, this object can act as a CORBA client for invoking other miscellaneous operations at the end-point with purposes other than controlling the stream.
- *Stream Adaptor,* which operates as a receiver/transmitter of Video/Audio frames over the network.

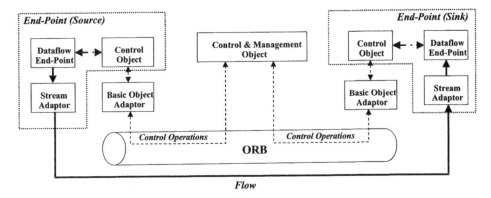

**Figure 4.7**  A simple streaming architecture

In addition to current specifications concerning the deployment of CORBA within a telecommunications environments, a number of other issues (such as RealTime-CORBA, wireless access networks, interworking with Intelligent Networks and Network Management Systems) are currently under specification. The impact the introduction of CORBA in these fields of telecommunications will have will be discussed in the following sections.

## 4.3.1  Network Management Systems

One of the first fields of telecommunications where CORBA became popular quite early was the Network Management area. Today's network management systems are mostly based on two widely known architectures:

- *Telecommunications Management Network* (TMN) coming from ITU-T (M3010, 1992), for operating, administrating and managing telecommunications networks. TMN contains standards interfaces and protocols, and aims to provide the appropriate management infrastructure for future telecommunication networks. The acknowledgement that the latter are going to evolve towards software-based networks (implying posing greater management requirements) provided the basis for the definition of this management architecture.
- *Simple Network Management Protocol* (SNMP) defined by IETF (RFC1098, 1989), for the management and administration of IP-based networks. SNMP adheres to a model of management stations and management agents performing operations on the actual network elements after receiving instructions from the management stations. Management agents can also report network element statistics back to the management stations when queried. SNMP then comprises the definition of the protocol used between management stations and agents that is used to communicate management information between them.

The adoption of CORBA in the management domain can act as a catalyst towards the integration of these two different approaches in network management. The idea is that the convergence of data and telecommunications networks pushes for a uniform way of managing both networks in an efficient way. Since the TMN and the SNMP based management systems are using different management models, one way for them to interwork is through the provision of CORBA management facilities capable of interoperating with both TMN and SNMP systems. In this way, as long as the appropriate interfaces have been defined, CORBA's modularity and implementation hiding properties can be employed.

Using this approach, for instance, a manager in the CORBA domain would be able to interact with an OSI or an SNMP agent as if it were in the CORBA domain, ideally without any knowledge about the domain of the target object. For a description of the OSI management framework refer to X.700 (1992). The main advantage of this approach is that the strength of distributed, object oriented architectures in providing simple, flexible and open APIs can enhance network management procedures if combined with the ubiquitousness of well known and standardized protocols such as the *Common Management Information Protocol* (CMIP) or SNMP. This can lead to a flexible and distributed network management architecture capable of interworking with legacy CMIP and/or SNMP based systems, and suitable for the provision of new information models. For instance, CORBA can interwork with management systems using ITU-T's Guidelines for the Definition of Managed Objects (GDMO) (X.722, 1992) or SNMP's Structure of Management Information (SMI) (RFC1902, 1996). Additionally, user demands for Web-based monitoring and management systems can be considered as another factor prompting the adoption of CORBA-based management, since CORBA is widely used for Web-based applications, and is even directly supported by commonly used browsers.

To enable the interworking between management systems based on different technologies, it is essential to be able to provide a mapping scheme between the relevant object models in order to introduce mechanisms for handling protocol and behavior conversions on the domain boundaries. The realization of mechanisms based on CORBA for the interworking of different management systems depends heavily upon the successful settlement of two major issues (OMG JIDM, 1998):

- The provision of a translation scheme between the different object models of both management architectures, also referred to as 'Specification Translation'. This means that translation algorithms for both TMN and SNMP based systems need to be defined. Therefore, translation from ASN.1 (X.208, 1988) to IDL, from GDMO to IDL, and vice versa, should be done for TMN. Similarly, translation from SNMP's SMI to IDL, and vice versa, for SNMP based systems.
- The provision of a dynamic conversion mechanism between the protocols and behaviors used in both domains, also referred to as 'Interaction Translation'. A set of CORBA facilities is required to support the interworking between different management environments. Two different interface levels can be identified for these facilities depending on the level of abstraction:
  - Management model independent interfaces. These facilities provide a generic framework to access a managed domain, independently from the management reference model that is being used.
  - Management model dependent interfaces. These facilities provide a lower level interface to existing information models of TMN and SNMP based systems. In this way, the generic interface is extended to support specific interactions between the two reference models.

In Figure 4.8 (a) a general diagram of a managing system comprising the manager and the managed system containing the agent is shown. In conventional systems the manager and the agent can be OSI based (TMN) or Internet based (SNMP), and the protocol could be CMIP or SNMP, respectively. Figure 4.8 (b) and (d) depict how the scheme is modified with the adoption of CORBA in the managing system. Figure 4.8 (b) presents a managing system that contains a CORBA-based manager, interacting (through the IDL interface) with a CORBA/CMIP gateway. In this scenario the OSI-based agent remains unchanged, therefore this approach enables the smooth transition from conventional systems to CORBA-based management systems. In Figure 4.8 (c) the opposite scenario is presented. The managing system is based on a conventional OSI manager, the managed system hosts a CORBA agent and the managed objects of the information model are specified with IDL and not with GDMO. The protocol is based on an IDL object definition, and therefore the existence of a CMIP/CORBA gateway in the managing system is essential. This approach is a step forward from the previous scenario, since it involves changes in both the managing and the managed system.

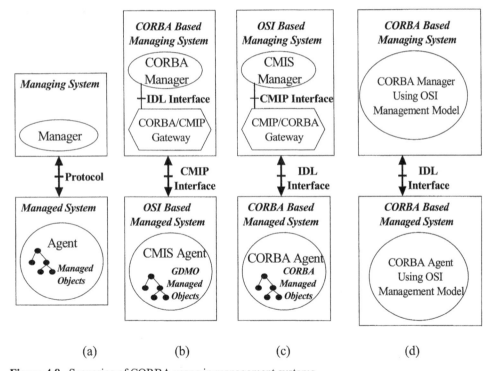

**Figure 4.8**  Scenarios of CORBA usage in management systems

The most radical scenario is depicted in Figure 4.8 (d), where both the managing and the managed system are based on CORBA. The only relationship with conventional systems in this approach is the preservation of the OSI management model for possible interoperability with OSI-based systems. The scenarios described above show the transition from a conventional OSI management system to a novel, CORBA-based one. The respective scenarios can be applied to SNMP-based systems as well.

### 4.3.2    Intelligent Networks

In *Intelligent Networks* (INs) (see Chapter 7), communication between network entities is based on the *Intelligent Network Application Protocol* (INAP). The transport of the INAP messages is accomplished using the *Transaction Capabilities Application Protocol* (TCAP), which is in turn based on the Common Channel *Signaling System 7* (SS7) protocol stack. TCAP is a transactions based protocol whose main facility is to carry out transactions between two communicating peers that are members of a dialogue.

In the case where a distributed computing infrastructure is laid between some or even all pairs of communicating nodes, direct CORBA method invocations can replace TCAP message exchange. In this case, each INAP information flow is transformed into a CORBA interface method bearing the same message and having the same arguments.

The use of CORBA as the communication-enabling medium has some interesting consequences:

- It introduces location transparencies into the environment.
- It facilitates seamless interworking with legacy code (implementation transparency).
- It calls for interworking provisions for equipment that uses traditional INAP.
- It allows software components built on any physical entity to leverage on available CORBA services. It also makes the boundaries between different physical entities less definitive.

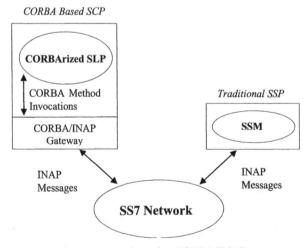

**Figure 4.9**  Interworking of CORBA/IN through a CORBA/INAP gateway

Location transparency coupled with the ability to use direct method invocations instead of resorting to complex communication protocols can have a significant impact on the maintainability of the code. The main advantages of DOT technologies, that of facilitating method invocations on a remotely located piece of code and, more importantly, their characteristic of allowing operations to be expressed using the same syntax as in the case of local method invocations, can be fully exploited in an IN environment where many dialogues are held between communicating peers. While this capability corresponds to non-functional aspects of programming and does not extend traditional IN capabilities from an operational point of view, it nevertheless gives the network designer greater flexibility. For instance, the boundaries between different physical entities are less definitive. In traditional IN, software residing in a *Service Switching Point* (SSP) could communicate with software residing in a *Service Control Point* (SCP) through the exchange of INAP messages. When both SSP and SCP are CORBA augmented, communication between software modules in SSP and SCP can extend what INAP information flows prescribe.

Current specifications for the interworking between classical IN systems and CORBA-based ones propose two different approaches (OMG Interworking, 1997). The first approach is based on the concept of a CORBA-to-INAP gateway. Figure 4.9 shows the architecture of an IN system employing this concept. For the realization of such an interworking architecture, the same issues discussed in Section 4.2.1 need to be addressed. As far as the translation scheme between the different models is concerned, the use of a translation algorithm (compiler) is necessary to cater for the translation between INAP constructs defined in ASN.1 and equivalent constructs defined in IDL. Regarding the dynamic conversion mechanism between the protocols, mapping mechanisms for the conversion of the IN-to-user interaction semantics need to be defined. In Figure 4.10 the *Service Logic Program* (SLP) and the *Switching State Machine* (SSM) that are the 'core parts' of the SCP and the SSP, respectively, are implemented as CORBA applications.

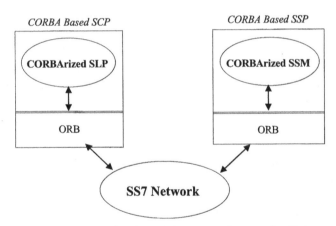

**Figure 4.10**   Interworking of CORBA/IN through GIOP mapping onto the SS7

The second type of interworking architecture (see Figure 4.10) considers the existing SS7 as the underlying transport infrastructure for the communication between CORBA-based IN entities (such as SCP and SSP). In this case, the existence of a GIOP to SS7 mapping is essential for the support of CORBA communication over SS7.

Even though the specification process has made some notable steps recently, there are still some open issues, mostly concerning the ORB's real-time capabilities, since the existence of this type of capabilities is a prerequisite to the commercial introduction of signaling and switching systems based on CORBA. It is imperative, for instance, that a real-time ORB to be used in such a system must support end-to-end predictability. End-to-end predictability means that thread priorities between client and server must be used in order to resolve resource contention during the processing of CORBA invocations.

Additionally, the duration of thread inversions during end-to-end processing along with the latencies of operation invocations must be bounded. Apart from the built-in capabilities that a real-time ORB must have, there are a number of external factors that need to be taken into consideration when designing an overall CORBA-based system with real-time capabilities. Namely, the available operating system's real-time support, the performance of the underlying communication protocols and the existence of optimized memory management routines are the main external factors that are crucial to the robust operation of a CORBA based real-time system. Currently, the first specifications for a real-time CORBA system (RT-CORBA) have been released (OMG Real-Time, 1999):

- *Real-Time CORBA Priority Mechanism.* RT-CORBA defines a general, platform independent priority scheme called Real-Time CORBA Priority. Essentially, this functions as a layer of abstraction, and has been introduced with the intention of hiding from the applications the differences in the priority schemes that different underlying operating systems may use.
- *Native Priority Mappings.* Since RT-CORBA uses an abstract priority layer, it is necessary to introduce a mechanism to map between this generic scheme and the OS-implemented native priority schemes. The `PriorityMapping` interface is defined with that purpose.
- *Thread-pools.* The RT-CORBA uses the thread-pool for managing the execution, on the server side, of the relevant threads. Thread pools are a widely used concept not only in middleware technologies but in multithreading applications in general. Essentially, the thread pool is a performance optimization mechanism that increases the design complexity of a system in order to make more sophisticated use of its threads. The thread pool in RT-CORBA offers a set of very useful features for thread management. The first feature is the pre-allocation of threads that allows the application programmer to ensure the existence of enough resources, in terms of threads, for serving concurrent invocations. In this way the programmer avoids the destruction and recreation of threads, which is a process that may by itself consume significant resources. The partitioning of threads is another feature which allows one part of the system to be isolated from the execution of another with potentially lower priority, since they are running on separate threads. In this way the overall performance of the system can be increased. Last but not least, the bounding of thread usage and the buffering of additional requests are key aspects of the thread-pool abstraction. By imposing limits on the number of threads a specific application may use, the system ensures that other parts of the system as well as user applications will not suffer from thread starvation. Moreover, when the maximum number of threads to be used is reached, incoming requests may be queued in appropriate buffering mechanisms.
- *Thread Scheduler.* The entities that need to be scheduled in the RT-CORBA system are the individual threads. In general, a thread represents a single flow of control (i.e. a sequence of instructions to be executed) within a machine. RT-CORBA specifies interfaces through which the characteristics of a thread that are of interest can be manipulated. These interfaces are the RT-CORBA `current` and `Threadpool` interfaces.
- *Real-Time CORBA Manager.* This entity is responsible for handling operations related to the configuration of the real-time ORB. It also manages the creation and destruction of objects implementing RT-CORBA IDL interfaces. For consistency, RT-CORBA

applications always use the generic CORBA priorities to express relative priorities in a system, even if all nodes are known to use the same native thread priority scheme.

### 4.3.3   Mobile/wireless networks

With the emergence of wireless network technologies – such as the *Global System for Mobile Telecommunications* (GSM), satellite telephony, wireless *Local Area Networks* (LAN), *General Packet Radio Service* (GPRS), *International Mobile Telecommunication* (IMT)-2000 / *Universal Mobile Telecommunication System* (UMTS) – it is important that CORBA is extended to accommodate networks that contain wireless links. Mobile devices typically have access to fewer local resources than stationary devices, since they have limited processing power and memory space. Consequently, the ability to use the services of remotely located objects becomes even more important for mobile devices. Even as these devices become more powerful and sophisticated, however, users will increasingly expect to be able to access services, that they currently use from stationary devices, from their mobile devices – raising expectations higher. If CORBA cannot properly support this, it will create a vacuum that some other technology (probably a domain-specific one) will attempt to fill.

The problem is that CORBA, as it exists today, is less than perfectly suited for wireless access. Certain solutions that might have been appropriate for wireless networks were not incorporated into CORBA, for fear that they may have adverse effects on applications using wired access. The fundamental challenges facing CORBA in wireless and mobile environments include providing reliability of transport (reliability is much harder to ensure in wireless networks than in fixed networks) and handling mobility of terminals where a terminal's point of attachment in the network may change.

The overall architecture of CORBA is sufficiently abstract and flexible to accommodate the differences in the characteristics of wireless networks with respect to wired ones, without requiring changes in its basic model. However, at the protocol and application levels, which current CORBA implementations require, changes will be unavoidable (Raatikainen, 1998).

Concerning the protocol level, latencies in wireless networks are generally greater than in fixed networks, throughput is generally lower and may vary unpredictably, and losses and errors are likely to be more frequent. As several studies have shown, TCP/IP performs poorly in these circumstances. One reason is that the higher latencies of wireless connections may cause a TCP/IP driver to assume that data has not been received correctly, and to needlessly retransmit, wasting bandwidth in the process. Another issue is that TCP/IP was designed with the assumption that errors are likely to result from network congestion, and has accordingly implemented flow control procedures using as feedback the occurrence of such errors. Consequently, TCP/IP drivers will inappropriately reduce transmission rates when they encounter errors. IIOP, on the other hand, foresees connections between objects that are maintained continuously. Wireless connections are more fragile than fixed connections and there are many circumstances in which it is not possible to keep a connection open continuously (for instance, when one uses a GSM device on a moving vehicle intermittent network coverage may be experienced). Wireless connections may also be charged on a time basis so that it may be prohibitive to keep connections open continuously. Ideally, it should be possible to wrap an intermittent communications channel inside an abstraction that handles disconnections and reconnections automatically. Moreover, IIOP is designed so that no state information is preserved between a client and a server object with TCP/IP connections. When a server becomes aware that a connection has been lost, it is expected to cancel any pending operations from clients of that connection. CORBA's Dynamic Invocation Interface provides a form of invocation called 'deferred synchronous' in which a client object can make a request with a non-blocking call and then poll for the result later. However, the actual remote invocation for a deferred synchronous call is synchronous, and the result is actually buffered on the client side. IIOP currently lacks a mechanism for allowing the server to buffer responses until clients are able to accept them. Such a mechanism would be useful in the case of an unreliable network.

At the application level, current CORBA applications running on fixed networks can safely assume that network disconnections are rare, and that when they do happen they can be considered exceptional enough to warrant active intervention by the user. Similar safe assumptions are that network latencies are low and that bandwidth is more or less fixed. However, on wireless networks, disconnections and reconnections may be of routine, and both latencies and bandwidths may fluctuate. To accommodate this broader range of circumstances gracefully, CORBA applications will need more sophisticated application-level facilities for handling them.

In OMG there is an ongoing work to identify the peculiarities of wireless networks as these are evidenced with respect to CORBA. The aim is to produce specifications that will provide sufficient support for the realization of distributed object applications for wireless networks. These specifications would be useful for both ORB vendors and developers of CORBA wireless networks' applications.

### 4.3.4   Telecommunications Information Networking Architecture

The *Telecommunications Information Networking Architecture* (TINA) defines a framework for the development of service and network management applications.

The TINA approach foresees that service components are deployed on a *Distributed Processing Environment* (DPE), which may in turn be implemented on top of different network infrastructures. That DPE is used for both the transmission of control information (remote method invocations) and the multimedia streams that may flow between user applications. Therefore, the network has to satisfy the connectivity requirements of different types of services, such as multimedia, multiparty or multicasting services. In an environment where the offered services are built by different, possibly unrelated vendors, care should also be taken to guard against the adverse effects of feature interaction. Finally, there is the need to interwork with different network technologies (Dupuy, 1995). It is not surprising, then, that TINA defines an elaborate management structure. Within TINA three different management levels are specified:

- The *Network Element (NE) Level* is concerned with handling the information that is required for managing specific equipment resources that provide NE-layer functions. This refers to the information required to manage the physical, telecommunications, and support functions within one NE.

- The *Resource Management Level* is concerned with the information representing the network, both physically and logically. This view is used to represent a global view of the network. It takes care of how individual NE entities are related, interconnected, and configured to provide and maintain end-to-end connectivity.
- The *Service Management Level* is concerned with the management of services, software and service configurations, and the utilization of resources necessary to provide the actual services. The use of resources by the service places requirements on the network's capabilities. Appropriate resource models are then necessary to provide for systematic responses of these issues.

There is a direct relationship between the three management levels mentioned above and the management levels of the TMN architecture. Specifically, the Service Management Level corresponds to the Service Management Layer of TMN, the Resource Management Level corresponds to the Network and Network Element Management Layers of TMN and the Network Element Level to the Network Element Layer of TMN.

From the computational point of view, TINA considers that the service and network management and control facilities are deployed on a distributed, object oriented processing environment, such as CORBA. This implies that interactions between computational objects follow the client-server model. To implement this model various relationships have been identified, and for this reason TINA defines the concept of *session* (Pavon, 1998). A session defines a relation between the members of a service, also known as *parties* or *resources*. Relationships between parties can be peer to peer, and comprise negotiation mechanisms. Relationships involving network resources follow the client-server model. TINA identifies three major types of sessions:

- *Access Session.* This type of session models the user actions for accessing a service. Examples of such actions can be considered the user authentication and authorization, the service customization, the user and terminal mobility handling.
- *Service Session.* This type of session represents the control and management actions needed for the support of a specific service from the system point of view (e.g. users involved, terminal capabilities, user profiles) and from the service point of view (e.g. type of communication between users, service functionality).
- *Communication Session.* This type of session encompasses the set of control and management procedures from the network's resource point of view in order to support the needs of services making active use of the underlying network infrastructure. From the TMN perspective, the Access and Service Session types would be considered as part of the Service Management Layer while the Communication Session type as a part of Network Management Layer.

The benefit of distinguishing between different types of sessions is that different procedures and mechanisms can be associated with each type independently from the others. This allows greater customization in response to an individual type's characteristics that may require special handling. Also, in this way the modification of a particular session model will not have a major (or any at all) impact on the modeling of other types of sessions involved in a specific service.

Current network management approaches provide efficient access methods based on the manager-agent paradigm. Contemporaneous approaches have not been evaluated for the open *Managed Objects* (MOs) access model proposed by CORBA. The concept of session and the way it is used by TINA may provide a solution to this. On the other hand, CORBA provides a great degree of flexibility in the distribution and implementation of managers and MOs. When a complete set of TMN like, management facilities become available from commercial CORBA platforms, management systems designers will be able to fully exploit the inherent advantages of CORBA in enriching the manager-agent model.

# References

Chung, P.E., Huang, Y. Yajnik, S., Liang, D., Shih, J.C., Wang, C.Y. and Wang, Y.M. (1998) DCOM and CORBA Side by Side, Step by Step, and Layer by Layer. *The C++ report*, January 1998.

Dupuy, F., Nilsson, C. and Inoue, Y. (1995) The TINA consortium: toward networking telecommunications information services. *IEEE Communications Magazine*, **33**(11), 78–83.

IETF RFC1098 (1989) Simple Network Management Protocol.

IETF RFC1902 (1996) Structure of Management Information for Version 2 of the Simple Network Management Protocol (SNMPv2).

ITU-T Recommendation M3010 (1992) Principles for a Telecommunications Management Network.

ITU-T Recommendation X.208 (1988) Abstract Syntax Notation One (ASN.1).

ITU-T Recommendation X.700 (1992) OSI Management Framework.

ITU-T Recommendation X.722 (1992) Guidelines for the Definition of Managed Objects (GDMO).

OMG, CORBA Asynchronous Messaging and QoS Service Control (1998), http://www.omg.org/cgi-bin/doc?orbos/98-05-05.

OMG, CORBA Component (1999), http://www.omg.org/cgi_bin/doc?orbos/99-02-05.

OMG, CORBA Facilities Architecture Specification (1998), http://www.omg.org/corba/cf2.html.

OMG, CORBA Interoperable Naming Service (1998), http://www.omg.org/cgi-bin/doc?orbos/98-10-11.

OMG, CORBA Objects-by-Value (1998), http://www.omg.org/cgi-bin/doc?orbos/98-10-06.

OMG, CORBA Scripting (1998), http://www.omg.org/cgi-bin/doc?orbos/98-12-08.

OMG, CORBA Services Specification (1998), http://www.omg.org/corba/sectran1.html .

OMG, CORBA Telecoms Specification (1998), http://www.omg.org/corba/ctfull.html.

OMG, CORBA/Firewall Security (1998), http://www.omg.org/cgi-bin/doc?orbos/98-05-04.

OMG, CORBA/IIOP 2.3 Specification (1998), http://www.omg.org/cgi-bin/doc?ptc/98-12-04.

OMG, Interworking between CORBA and TC Systems (1997), http://www.omg.org/docs/-telecom/97-12-06.pdf.

OMG, Java Language to IDL Mapping (1998), http://www.omg.org/cgi-bin/doc?orbos/98-04-04.

OMG, JIDM Interaction Translation (1998), http://www.omg.org/docs/telecom/98-10-10.

OMG, Minimum CORBA (1998), http://www.omg.org/cgi-bin/doc?orbos/98-08-04.

OMG, Real-Time CORBA (1999), http://www.omg.org/cgi-bin/doc?orbos/99-02-12.

Orfali, R., Harkey, D. and Edwards J. (1996) *The Essential Distributed Object Survival Guide*. John Wiley & Sons.

Pavon, J., Tomas, J., Bardout, Y. and Hauw L.-H. (1998) CORBA for network and service management in the TINA framework. *IEEE Communications Magazine*, **36**(3), 72–79.

Raatikainen K. et al. (1998) White Paper on Wireless Access and Terminal Mobility in CORBA. OMG Doc: telecom/98-11-09.

Redlich, J.-P., Suzuki, M. and Weinstein, S. (1998) Distributed object technology for networking. *IEEE Communications Magazine*, **36**(10), 100–111.

Tallman, O. and Kain J.B. (1998) COM versus CORBA: A Decision Framework, Distributed Computing, September-December, http://www.quoinic.com.

Yee A. (1999) Making Sense Of The COM versus CORBA Debate. http://www.-performancecomputing.com/features/9906dev.shtml.

# 5

# Machine Independent Code

## 5.1 Introduction

By Machine Independent Code we refer to programs that can execute – without requiring recompilation – in a variety of machines irrespective of underlying Operating System (OS) and/or hardware architectures. In the early days of programming, where implementation was carried out in machine or assembly language, there could be no such concept since there was no compiling. A program was written specifically with a certain machine in mind, and the correspondence was so strong that modifications were needed even for a new version of a processor.

Once the first programming languages were invented, the importance of the ability to write a program that could run on different machines without modifications was quickly recognized. All the more so, as software development costs began to increase exponentially and soon dwarfed those of hardware equipment. The proliferation of operating systems and hardware platforms meant that commercial software often had to be ported in an architecture different than that for which it was original developed. Source code portability became a major concern. Intertwined with source portability is the used compiler technology. Source code is after all input to a compiler, and it is the ability of different compilers to generate code for different systems using the same sources that determines the portability of a certain language. The ability to write a program as a set of co-operating functions or classes gave rise to the implementation of standard libraries that encompassed architecture related facilities which would require modifications in case porting to a different system was considered. A program making use of these libraries was much easier to port to a different system once the respective libraries had themselves been ported.

As new languages were defined, compilers and/or interpreters for them were also developed. Any given language can lend itself either to compiler or interpreter implementation. For instance, the original implementation of the BASIC programming language was a compiler. In subsequent years, interpreters for BASIC became (and still are) very popular. It can be seen that compiler or interpreter implementation is rarely something that derives from the definition of the language. For the most part, it is linked to considerations over the speed at which programs written in that language can execute, or their portability. Therefore, when talking about language portability we actually refer to its

*Object Oriented Software Technologies in Telecommunications*, Edited by I. Venieris, F. Zizza and T. Magedanz
© 2000 John Wiley & Sons Ltd

implementation. Interpreted versions of programming languages can easily become machine independent. Once the interpreter is ported to different platforms and preserves the same specifications for the language definition, the source code of every program can be ported among those platforms without modifications. In addition, since the source code of a program written in an interpreted language usually represents its final interchange form, then the language itself can be considered as being machine independent. On the other hand, the compiled languages produce binary executable files, which constitute the final interchange form of a program written to that language. By definition, this final form is heavily machine dependent: the binary file can be executed only by the specific machine that it has been produced for, and uses native OS facilities only. The format of the binary would be different if a compiler for a different machine or OS was used. Therefore, there is no portability in the final output of compilers. This is not to say that portability is of no concern for compiled languages. While compiled languages can have no binary portability, there is still room for source code portability. Source code portability or source code machine independence refers to the ability of the same program to be compilable under different processors and operating systems. Source portability can be impaired if a program relies on operating system calls that – as is usually the case – do not have counterparts on other systems, or if it uses libraries that are operating system specific. Since the role of the operating system has grown considerably in importance compared to the early days of programming when the OSes were either very thin or circumvented by assembly-level programmers, it is very easy to destroy the portability of a body of code across different architectures. All the more so if a program makes extensive use of OS-related functionality such as Graphical User Interfaces (GUIs). As we have noted, the definition of standard libraries for most commercial languages has helped ameliorate the situation. A program that rests only on these libraries can be easily ported to different machines requiring little or no modifications at all. The libraries are interfacing with the actual OS functionality and are implemented differently for each architecture, but their interface with the programs that use them is the same. The standard C library is a good example. Unfortunately, few programs can rest exclusively on standardized libraries. Implementation of graphical user interfaces has to rely on the native OS. Therefore, it looks like machine independence is very hard to achieve even at the level of the source code. The proliferation of different versions of the most popular OSes and the increased reliance of programs on OS characteristics exacerbate this problem, which is further precipitated by questionable programming practices.

It is possible – and in fact a considerable amount of effort has been invested in this – to construct a program in such a form as to allow the bulk of the code to be platform-independent. In this case, functions or objects that by necessity have to interact with the operating system or that cater for machine dependant considerations can be located in different modules. In the case where this code will be used in a different environment, only these specific modules will need to be modified. This is not different to the approach taken in the definition and implementation of standard libraries for each language.

In any case preserving – much more guaranteeing – code portability is a tedious task that burdens the programmer and is not offered directly at the language level. The interpreted languages, as we have noted, are an exception to this rule, but their use is associated with significant performance degradation. Another common problem of interpreters is that usually the largest portion of interpreter-detectable mistakes are shown up during runtime, while a strongly typed language and a good compiler implementation can help detect a large number of common errors at compile time thus leading to a more efficient code-compile-debug cycle.

The advent of Java created new potential for writing machine independent software by tackling the problem at the language level (instead of deferring it to the programmer).

## 5.2   Java

### 5.2.1   Introduction

Java is an object oriented programming language. Its syntax is reminiscent to that of C++, but the language differs substantially in that all error-prone, esoteric or exotic features of C++ have been left out. Being a much simpler and orthogonal language (usually allowing only one way of doing something), it eliminates common pitfalls and sources of bugs and allows the developer to be more concerned with the actual problem he faces. Java, for instance, eliminates pointer arithmetic, and in doing so also eliminates a large source of hard to find runtime errors. Similarly, it encompasses a garbage collection mechanism that exempts the developers from the responsibility of explicitly de-allocating memory. An insightful book covering the most essential features of Java from a designer's point of view is that by Eckel (1998).

Java is also a much 'cleaner' language compared to C++. Whereas C++ is a hybrid object oriented, structured language, Java adopts a pure object oriented approach. It does not, for instance, allow global functions or data, and mandates that all classes should ultimately descend from a common base class (called Object). Java also supports only single inheritance, as do many other object oriented languages. C++ had to provide for backward compatibility with C, which burdened it with a host of unnecessary, redundant features. Java, in contrast, suffered from no such impediment. Academic notions of purity, however, were not pursued to the detriment of the practicality of the language or of its simplicity. Java remains a conceptually straightforward, elegant language that can be used by anyone with a modicum of understanding of object oriented concepts.

These characteristics by themselves would allow Java to at least enjoy a certain degree of popularity. It was the fact that the language had the fortune to ride on the Internet wave that ordained its persuasiveness. In fact, Java became so immensely popular with the Internet that a common misconception was that it was mainly a Web technology comparable, for instance, with JavaScript (the name similarity did little to correct this impression). It took a while to become apparent that Java is a fully-featured language capable of a lot more than contributing to fancy web pages. Nowadays, Java has entered the mainstream and is used for demanding, commercial applications. Over the years it has significantly eroded the C++ software base, and will undoubtedly continue to do so in the future. The characteristic of Java that was most lauded, and continues to exert the greater influence in the growth of the language's share is its inherent machine independence. How is that guaranteed?

Let's start by saying that Java is a semi-interpreted language. This means that both compilation and interpretation are involved in the process of writing and executing a Java program. The Java compiler is different from traditional compilers in that it does not produce object code for an actual hardware implementation, but rather for a virtual machine. This machine is called Java Virtual Machine (JVM), and it resembles a specification for an actual processor. Though JVM can be directly compiled to silicon or be implemented at microcode level, commercial JVMs have software implementations that act as emulators for the specified CPU. The JVM specification defines the set of available registers, the instruction set and other common memory areas such as the stack and the heap. It also mandates certain procedural aspects of the machine's operation such as the existence of a garbage collection mechanism and linking procedures. Many of these aspects pertain to an operating system's implementation and not to a microprocessor design. The JVM therefore prescribes a standard way of carrying out certain common OS facilities, as well as providing an Application Programming Interface (API) to these, in this way resembling an Operating System as well as an actual processor. You can think of the JVM as a 'high level processor'. For instance, its instruction set is not a common one since it incorporates object oriented concepts right into the heart of it!

Since Java 'executables' are meant to be executed on the JVM, the binary files that are produced by the Java compiler are meaningless to any concrete processor, and are referred to as *object bytecodes*. Their name is actually a hint at the common opcode length which is only one byte (resulting in small 'executables'). What is important, however, is that this step of compilation means that the original source code has been transformed into something of a lower level that can more easily be on-the-fly interpreted. In fact all interpreters one way or another include a 'compilation-like' phase. An intermediate form is usually produced, and it is this form that is interpreted immediately afterwards. The difference in Java is that a specification exists for this intermediate form (the class format defined in the JVM specification), and that this form (the object bytecodes) are not produced each time the program is executed but only once during compilation. The developer writes the program, 'compiles' it, and can then treat the bytecodes as the 'executable'. For every machine that there is an implementation of the JVM available, these bytecodes can be interpreted at speeds roughly comparable to that of native code. The JVM is implemented using native code at each system, and is essentially the interpreter. Running a Java program at a new system simply requires porting the JVM for that system. In that respect, JVM can be viewed as acting as an intermediate between the Java bytecodes (which are compiled irrespectively of the targeted platform) and the actual hardware and OS resources (see Figure 5.1).

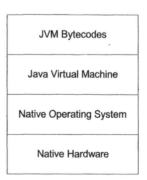

**Figure 5.1**  The JVM acting as an intermediate

It is worth noting that there is nothing inherent in the Java language specification prohibiting a compiler implementation. Again, there are tools that can produce an executable as well as bytecodes. Furthermore, compilers can be implemented for other languages to produce not native code but bytecodes. These bytecodes could then be executed on any JVM implementation. Ada is a typical example, since its architecture is particularly well suited to producing different code formats. Also, concrete implementations of the Java VM (referred to as Java chips) can be produced allowing the bytecodes to be executed in an actual processor, and not a software-emulated virtual machine.

Up until now, Java looks like a hybrid. It incorporates elements of a compiled and an interpreted language. The interpreted part clearly enhances its portability across different platforms. It cannot, however, by itself explain how operating system dependencies are not allowed to creep in and mar its machine independence. The reason is that Java is not only a language: it also comprises aspects of an operating system. This needs a little elaboration.

The concept of an operating system is two-fold: it can refer to the runtime environment that is responsible for such tasks as scheduling, maintaining a consistent file system state or providing the illusion of an unlimited virtual memory. It can also, however, refer to the Application Programming Interface that is visible to the programmer and gives him access to such aspects of the machine's operations as the Graphical User Interface, file handling and memory management. Part of the rationale behind the definition of standard libraries is to ensure that most of these tasks are appropriately implemented in separate libraries. These libraries can then be ported in different operating systems, while programs that rely on their standardized APIs do not need to change since they are not themselves interfacing directly with the operating system. An essential part of the 'write once run everywhere' characteristic of Java is the fact that it comes with standard class libraries covering all OS-dependant areas of programming a developer might get involved with.

It is important to understand the different effect that the standard class library and the virtual machine concept have on the portability characteristics of Java. The definition of standard libraries ensures source code portability across different operating systems. The definition and enforcement of the virtual machine architecture ensures binary (object-code) portability. So, a compiled program in Java can run as it is, on different hardware or OS platforms without even requiring re-compilation. The only requirement is that the Java VM itself has been ported to that platform. Porting of the JVM requires porting of the relevant, OS-related libraries as well. In this sense Java is something more than a language, it is a platform or an operating system – under the second, narrower interpretation of the term.

Other languages also come with standard libraries (e.g. C, Ada). These libraries, however, either do not cover the entire spectrum of functions available to the programmer in a modern GUI-based operating system, or if they do, the respective language does not provide for a binary standard, in this way reducing the effectiveness of the libraries to source code portability. In effect the JVM specification uses the abstraction of an actual hardware processor to define a vocabulary, which it uses to specify in a concise way the code's binary format. While source code portability is by itself very important, without binary code compatibility, certain innovations that were largely based on Java (or in any case preceded it as concepts but remain largely latent until its wide acceptance), would be much harder to implement. The ubiquitous web applets and Mobile Agent Technology are fitting examples.

## 5.2.2    The Java Virtual Machine architecture

As we have mentioned, the JVM defines an abstract computing machine which can 'execute' the bytecodes that are the product of a Java program's compilation. The JVM, however, has nothing to do with the Java language *per se*. It is unconcerned about its semantics, its concepts, and its grammar. Like a real processor it is designed to execute a specified instruction set. Java programs are compiled into this instruction set, but the same can happen (and in fact, does happen) for programs written in other languages as well. The idea of a Virtual Machine is a powerful concept and one that was formulated long before the invention of Java. Historically, the term 'Virtual Machine' was coined in the context of Operating Systems to refer to an operating system's abstraction that presented each running program with a separate, logical view of the machine in spite of the fact that all programs run on the same hardware node. In this way, a separate, 'virtual' CPU is available to each process. This prevented malicious or malfunctioning programs from crashing the entire machine and offered a clean-cut way to ensure that no program will write on data kept by an other running program (since each process was offered its own virtual address space as well). The notion of an operating system implemented with a virtual machine architecture is thus juxtaposed against other OS implementation strategies such as the layered or micro-kernel architectures. Correspondingly, this term has been used by many operating system releases to refer roughly to this sort of facility.

This concept is employed in Java with a different purpose. It is not used as an operating system's design strategy, but rather as a way to implement a language. This is not unprecedented either. For instance, the P-System developed at the University of California at San Diego (UCSD) was a portable operating system that like modern day Java was based on a virtual machine with a standard instruction set (*p-code* instructions) which were emulated on different hardware. Pascal compilers that produced p-code instructions could be used to compile programs that ran under Apple, Xerox or DEC machines in much the same way that Java compilers produce programs that can run under Solaris or Windows. Unsurprisingly, the most popular language for the P-System was UCSD Pascal, just as the most popular language for writing programs to be executed on the JVM is Java. Nothing prevented compilers of other languages from generating p-code instructions, at least no more than todays' compilers for C++ are prevented from generating bytecodes.

Perhaps the most important part of the JVM specification is the definition of a class file format. To a JVM implementation a class is a sequence of bytecodes that represent instructions. A JVM is expected to read this bytestream and correctly perform whatever operations are prescribed therein. When the JVM starts, it needs to execute a certain class. There is nothing in the specification that mandates that this class must be stored in a file. The bytestream can be computed by the machine itself on the fly, fetched over the network or fed to it be another entity. The 'executing' modules of JVM do not distinguish how the bytecodes were produced or stored. It is the task of a *class loader* to load them into the JVM. We will talk about the classloader later, but it is necessary to underline its importance, since this ability to introduce classes into the JVM in a variety of ways and in a manner transparent to the computing modules opened the way for applications such as Web applets and Mobile Agents. In the typical case, however, the bytecodes are produced offline by a compiler and are stored in a typical file (with certain name conventions applying). The class to execute at startup is then provided as an input argument at the invocation of the JVM from a command line prompt.

The class format is specified in detail and is hardware and operating system independent. It also determines byte-ordering ensuring that the same bytecodes run on both big and little endian systems. What makes the defined by the class format instruction set an unusual one is that it includes reference types in addition to primitive ones. Most instruction sets usually define a set of primitive types and numeric operations on them. They also define a 'function call' such as the CALL instruction in the 80x86 family of processors. The JVM, however, includes object orientation right into the definition of its instruction set by including reference types, in addition to primitive ones and supporting the notion of 'invoking a method on an object' instead of simply 'calling a sub-routine'. Except for that, bytecode instructions look like normal instructions of a typical instruction set: mnemonics may be defined for each machine instruction and operands are specified to the instructions in the usual ways employed by 'real' assembly languages. For instance, Figure 5.2 shows two instructions with different number of operands.

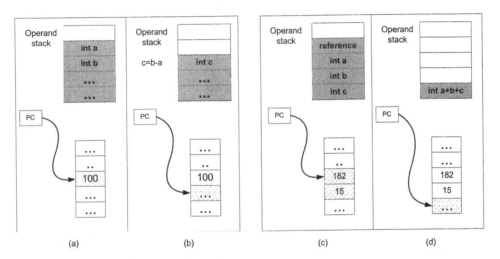

**Figure 5.2**   Execution of two JVM instructions

The first one, (a) and (b), corresponds to the instruction used to subtract two integers. The operand stack is implicitly used so this instruction takes no argument, only its opcode. Two integers are popped from the stack and the first is subtracted from the second. The result is then pushed back in to the stack. The operand stack memory areas are described below.

The second instruction, (c) and (d), corresponds to the invocation of a method using an object reference. In the JVM's memory layout, after the invocation instruction's operand code (decimal 182), an index to an entry in the constant pool (here, decimal 15) specifying the object's class, and the method to be invoked is found. In the operand stack, the reference to the method is stored first followed by the arguments to be supplied to the method. After the execution of the method, the result is pushed in the stack. The method's reference points to a symbol table area where information about a method's signature is kept. This allows the runtime implementation to ensure type compatibility between the arguments defined in a method's signature and the type of those actually offered at execution time.

Except for opcodes and operand addressing schemes for the instruction set defined, the

JVM specification goes on to describe several virtual 'parts' of the machine. These are of a more tangible nature than the instructions set architecture. In a real machine, some would correspond to an actual hardware implementation while others would be implemented in software at the operating system level. According to the JVM specification (Lindholm, 1996) these are as detailed below.

The PC register implements the same functionality as that of a typical Program Counter in an actual processor. It stores the address of the next instruction to be fetched and executed. As the PC keeps track of the position in memory where instructions are executed, it is incremented by the JVM after the completion of each instruction to point to the immediate next one, unless a jump was instructed. Since a JVM may have multiple executing threads at once, each thread has its own PC.

Inside a single thread and each time a method is invoked, a *frame* is associated with it. Frames reflect the current environment, into which a method is invoked and operates, and contain among other things local variables and a data area referred to as an *operand stack*. The operand stack holds values of primitive types which are manipulated by arithmetic or logical instructions, popping values out of the stack and pushing results back in it, as shown in the two example operations described in Figure 5.2. It is used for holding intermediate results. Each running thread has at any time a single executing method, and the frame of that method is called the current frame. As methods return or invoke other methods on their behalf, the current frame changes. Frames of different threads are, however, kept separate. As the current frame contains all the necessary information to reproduce the state of the running method, a simple push algorithm is sufficient. As methods are invoked, new frames are pushed on top of current ones, while as methods return, frames are popped. Since a frame is related to an *individual* invocation of a method, recursion is readily supported. There are also three more frame-specific registers pointing to various data areas of the stack frame of the currently executing method. These are the *optop*, *frame* and *vars* registers.

On a thread level, a special memory area is assigned to each thread at the time it is created. This is the *stack* into which *frames* are allocated and reclaimed as methods are invoked and return in that thread. As we noted, individual frames are only pushed into the stack or popped out of it, so a simple heap allocation is sufficient for the implementation of the stack. Stacks and frames in typical programming languages such as C operate in much the same way. The only difference is that stack manipulation and frame insertion/retraction are implemented differently for each language depending on the compiler. Stack and frame management is specified by the JVM definition itself irrespective of the initial language for which bytecodes were produced. This removes the need for explicitly indicating calling conventions at the language level.

The *heap* is a specially designated runtime data area that is commonly accessed by all running threads. When objects or arrays are created, they are allocated on the heap. Java always uses the heap for non-primitive objects. The Java language provides no facility for explicitly freeing memory (occupied by an object or reserved for an array) and accordingly, JVM implements a *garbage collector* responsible for reclaiming heap memory, once respective objects or arrays become unreachable. Certain classes of 'reachability' are defined (unreachability being one of them), but no garbage collection algorithm is specified. It is left to the discretion of the implementors of the JVM to define appropriate algorithms and policies for reclaiming memory. The set of reachability classes was further extended with the definition of *weak references* in the Java 2 platform. Weak references are intended as a way to cache information in memory and maintain pointers to it while at the same time also allowing relative objects to be garbage collected when memory is low.

The heap also contains the *local method area* which stores among other things the actual method code, and the *runtime constant pool* which stores compile time constants or other types of constants that are only resolved at runtime.

### 5.2.3 The class loader

Of particular importance to the capabilities offered by Java (or more precisely, by the bytecode format and the JVM architecture) is the definition of the *class loader* object and the latitude that is given to individual programmers to extend it and enhance its capabilities.

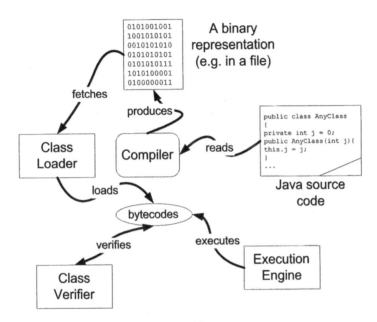

**Figure 5.3** The role of JVM's class loader and verifier

Figure 5.3 provides a graphical, high-level representation of the steps taken in compiling and executing a Java program and the place where the class loader fits the whole picture.

The JVM begins execution by entering a special method of a single class. This method must be *static*, meaning that it is associated with the class itself and not with any individual object of it, *final*, meaning that it cannot be overridden and must have a certain, very specific signature. As the JVM attempts to execute this method, it has to load first the code of the class in which this method is defined. Essentially, a byte stream in the class format corresponding to this class is required. The executing modules (the interpreter's core) are indifferent to the mechanism by which this stream will be provided to them. The entity responsible for providing this stream is the VM's class loader. In a typical stand-alone application the class loader will find the bytecodes stored in a file somewhere in the local file system. In the case of a Web browser executing an applet, the class loader of the built-in JVM will fetch the bytecodes for the applet using a *Hyper Text Transfer Protocol* (HTTP) connection. For the Mobile Agent Platforms described in the next chapter the class loader fetches the bytecodes from the platform where the incoming agent is migrating from.

As the first class is loaded, bytecodes for additional classes may be required, spanning more class loading. This process is termed *linking* and involves *verification* of the loaded bytecodes to determine that they correspond to a valid instruction set (for instance, that they adhere to the Class format), and that they will not pose any danger to the integrity of the machine in which they will execute (e.g. by ensuring that all 'jump' instructions point to valid addresses). The specification does not mandate when verification takes place, but the usual approach is to perform most verification tests immediately after the bytecodes are loaded.

The point in time when loading and linking occurs is again left to the implementor to decide. For instance, according to the early-linking strategy, required classes are loaded once the class that uses them is loaded. According to the late-linking strategy, classes are loaded when an actual instruction that uses them is encountered. To the implementation is also left the issue of how the class to be loaded at bootstrap is specified. For example, the Java 2 platform uses a different convention for that than the method used in Java 1.0 and 1.1.

The important point here is that while the JVM starts with a bootstrap class loader, the programmer can indicate his own class loader to be installed. In such a case, the bootstrap class loader will delegate responsibility to the programmer-defined one. While the bootstrap class loader is part of the machine's implementation, programmer-derived class loaders are created and treated just like any other object implemented by the programmer. This flexibility allows a programmer to adapt the class loading abilities of the JVM to his given circumstances, and implement novel ways of finding and loading classes. Since loading classes at runtime essentially corresponds to changing a program while it runs, this allows Java programs to exhibit a very dynamic behavior at runtime. A programmer has simply to extend a base `ClassLoader` class to define his or her own class loader in exactly the same way in which he would extend any other third-party library class.

### 5.2.4    Java in telecommunication protocols

Java can be used to implement a telecommunication protocol just like any other general purpose programming language. In fact in its early conception, Java purported to find a niche in the market for set top boxes. The idea was that Java coupled with the *Multimedia Hypermedia Experts Group* (MHEG) standard (see Section 5.4) would be used in multimedia presentations offered on typical consumer TV monitors augmented with a set top box. Both Java and MHEG are machine independent, so the relative code could be downloaded as appropriate in a set top box without hardware or operating system considerations. For this reason, a large number of bitwise operators were included in the language, as Java programs were expected to parse headers and do a lot of low level bit manipulation. Most of these operators have now fallen into obscurity, and are scarcely used, but they are still there and can be employed when an implementation requires so.

A protocol (or any other kind of software, for that matter) will be certainly implemented much faster in Java but will be executed slower. The simplicity of the language, the facilities it offers (e.g. automatic garbage collection) and the lack of pointer arithmetic (which is commonly responsible for many bugs) contribute to faster development. The semi-interpreted form in which programs of the language are usually executed is responsible for the inferior performance.

It is a strategic decision to estimate the performance penalty of interpretation, weigh it against the importance of speedier development, less maintenance effort as well as complete eradication of the need for porting and assess if the choice of Java as the implementation language is justified. Let's assume that in light of these considerations, an implementation in Java has been decided. What are then the peculiar properties of Java and how do they manifest themselves in telecommunication programs?

A common characteristic of protocol software traditionally implemented in C or in C++ is the heavy use of pointers. Processing or parsing a header mainly involves progressing a pointer through it, using pointer arithmetic. Packets flowing through different protocol layers or processed by different modules correspond to a certain pointer's referencing and dereferencing. Basically, copy operations are kept to a bare minimum, and information flows are based on pointer manipulation. There is nothing inherent in Java that prohibits this programming style. Java pointers are called 'references' and for the purpose of passing handles to data around (instead of copying it), they have the same functionality as pointers. Actually, all objects in Java are passed by reference minimizing further needless copying. Destructors are not called by the programmer but implicitly by the garbage collector. Memory allocated for arrays is similarly automatically reclaimed. Creation of objects in the heap and the execution of garbage collection mechanisms involve a mild performance penalty, but do not constraint pointer functionality in any way. What Java does not have, however, is pointer arithmetic. That is, it is not possible to increment or decrement a pointer using simple arithmetic operations. Also, the Java compiler is much stricter when it comes to casting from one pointer type to another than is, for instance, the C compiler. This precludes certain programming idioms, more importantly with respect to headers, and may necessitate certain copy operations to implement the needed functionality (Krupczak, 1998). The absence of the ubiquitous `#define` directives is not restricting since all necessary constants can be declared as public final fields in a single class. This deviates from programming practices that are deeply entrenched, but in fact corresponds to sounder programming as it makes more type information available to the compiler. Stroutstrap[1] himself advises against use of `#define` directives or in general, uncritical use of macros (Stroutstrap, 1996). Furthermore, in contrast to the `#define` directives, constants defined this way can change without requiring re-compilation of other classes that use them. In case minimizing access time to commonly accessed constants is important, then those fields can be declared as `final` becoming in this way compile time constants. This corresponds to forfeiting the advantage of easy changes of constant definitions to favor increased performance. Pre-processing and use of macros is also widely employed in typical protocol implementation. Java does not support macros, though strictly speaking, macros are not part of the language, since they are not observed by the compiler. While nothing precludes use of, for instance, the standard C++ pre-processor to resolve macros in a Java program before handing it for compilation, this practice would seem unnatural. Java emphasizes readability, and macros are notorious for obscuring an understanding of the program's behavior. They also make discovery of compile or runtime errors more problematic. When considering low level code, such as in protocol implementations, Java by not using macros, makes debugging easier, since type safety of expressions is guaranteed by the compiler (in contrast preprocessing is a textual modification procedure that replaces macro names with

---

[1] B. Stroutstrap is the designer of the C++ programming language.

literals – providing no type information to the compiler). Also, by not using macros, what the developer sees is what the compiler sees – there is no preprocessing involved between the two.

Overall, use of Java for implementing telecommunication software may discourage or prohibit employment of certain common programming idioms, but results in far more readable and modifiable source code. Performance of the resultant code will be inferior to that of natively compiled one only if bytecodes are used. Commercial, native compilers for Java are available, and in any case, the technology of JVMs is improving rapidly. Also, silicon implementations of JVM are forthcoming.

## 5.3    Scripting Languages

Scripting languages are in certain aspects similar to interpreted ones. For one, they are inherently machine independent since they are always interpreted. Their basic differences are that scripting languages represent a somewhat higher level of coding abstraction and are usually much more loosely typed. The line between 'normal' interpreted languages and scripted ones is therefore blurred. Another distinguishing characteristic is that the latter are also used for assembling 'components' together where these components are implemented in a more concrete language. Scripts are in this sense the glue that keeps pre-fabricated programming modules together or defines their interactions. It is also the case that scripting languages have a higher degree of specialization than general purpose interpreted ones do, which seek to be more generic.

Scripting languages' machine independence has made them very popular in the Internet. They are used for instance at Web pages to allow a more dynamic content. They are also used behind the scenes at the Web servers themselves for processing queries or input. In this latter case they take advantage of the definition of the *Common Gateway Interface* (CGI) that allows programs to be invoked from Web servers. Java can also be put in both of these uses in the form of applets or servlets.

## 5.4    The Standard for Coding Multimedia Presentations – MHEG

In Section 5.2 we examined the fundamental properties of the most widespread machine-independent programming language, Java. While Java is the leader (one could say that Java is the *defacto* standard) among general purpose, interpreter-like object oriented programming languages, MHEG (Meyer, 1995) is a special purpose (in the multimedia domain) object oriented application format. MHEG could also be considered as a programming language, since it is used to build applications. However, it is more an application interchange format with object oriented characteristics. It is illustrated in this book because MHEG follows a different paradigm than traditional programming languages do in order to bring object oriented design discipline in the world of telecommunication services. On the other hand, MHEG is a very appropriate example of building a specialized object oriented framework for supporting telecom services with a great level of programmability, efficiency and flexibility.

Offering interactive multimedia applications over *Wide Area Networks* (WAN) in an efficient manner is a difficult challenge for modern telecommunication networks. The reason is that this sort of service poses significant demands on the transport network in terms of bandwidth as well as *Quality of Service* (QoS) characteristics. For instance, a *Movies on Demand* (MoD) service requires the transport of digital video in real-time, which calls for a significant bandwidth allocation in order to reduce delay and jitter.

Furthermore, data sources of this kind are hardly predictable with regard to their bits per second (bps) information rate. This exacerbates even more the need for wider latitude over the bandwidth reserved at each specific point by the transport network. It can be seen, therefore, that digital multimedia services are especially demanding in relation to the actual capabilities of modern WANs. The latter are in comparison becoming less fit to cater for the stringent QoS requirements of the more advanced services. In the Internet, for instance, where network capabilities are limited in terms of QoS, there is an offering of multimedia services, albeit with a low quality. The inability of the present Internet to offer QoS characteristics and provide for them at the protocol level is one reason for the small growth of this set of services. The plethora of media formats and the diversity of hardware/software equipment available at the end systems that are connected to the Internet only precipitates the problem. A machine independent platform coupled with a standardized presentation format would deal with some of these problems.

MHEG (Bitzer, 1998) was one of the most important technologies invented for the implementation of such a platform. This technology is standardized by the International Organization for Standardization (ISO) by the same working group responsible for the video coding formats (*Moving Picture Experts Group* – MPEG series). The MHEG standard consists of a series of recommendations which, taken together, comprise a machine-independent 'language' for pre-orchestrating interactive multimedia presentations. More specifically, the MHEG series of standards comprises:

- *MHEG-1 (MHEG Object Representation, Base Notation ASN.1)*. This part, the first that was created, described a very rough mechanism of encoding multimedia applications. It is of little concern to implementors.
- *MHEG-2 (Alternate object representation SGML)*. This part of specification was intended to define a *Standard Generalized Markup Language* (SGML) for describing and encoding the mechanism defined in part 1. Nevertheless, this part of the specifications was never brought to completion.
- *MHEG-3 (MHEG Script Interchange Representation)*. MHEG-3 describes a scripting language called SCRIPT that would be used by objects defined in the first part (MHEG-1).
- *MHEG-4 (MHEG Registration Procedures)*. It defines a standard way of inserting indices to the various content formats that are defined in other MHEG parts.
- *MHEG-5 (Support for Base-Level Interactive Applications)*. This part details the general MHEG-1 specification having in mind low cost systems with minimal resources. This is actually the most important part of the specification from a commercial point of view.
- *MHEG-6 (Support for Enhanced Interactive Applications)*. MHEG-6 defines a mechanism for exchanging information, as well as an Application Programming Interface between MHEG-5 objects and the Java Virtual Machine. In this way, the interchange of Java object code wrapped within MHEG-5 objects is facilitated.
- *MHEG-7 (Interoperability and Conformance Testing)*. This part specifies conformance testing procedures of commercial MHEG implementations with the actual specifications.

## 5.4.1    MHEG-5

The heart of the specifications is part 5 (ISO, 1998). It specifies the final form representation of the software objects that constitute a multimedia application. Let's take a closer look at the basic components of a presentation-application encoded according to the MHEG-5 format.

The coding mechanism for MHEG presentations is conceptually based on object oriented programming methodologies. MHEG encoded presentation consists of objects produced out of a class hierarchy. These classes represent those elements (visual, interactive, associative) that can be composed to codify the interaction with an end user. There are classes representing the following categories of elements:

- Visual elements (e.g. geometric drawings, digital images, text, colors, color palettes)
- Multimedia elements (e.g. video, real-time graphics, sounds)
- Interactive elements (e.g. hypertext, entry field, button, check box)
- Structures organizing other elements (e.g. check lists, focusing format)
- Synchronization elements (e.g. synchronization of video with sound)
- Grouping elements (e.g. application, scene).

Combinations between objects containing these elements produce an organized informational structure, which is processed by a terminal to produce the actual interactive multimedia presentation. The way in which this processing takes place is explained in Section 5.4.6. The distinctive feature of MHEG is that it separates the active content (programming language type instructions) from the passive one (attributes), in contrast to usual interpreted or scripting languages. In this way, presentation data transmitted over the network, from the service provider to the actual user, need to comprise only the passive part of the presentation. A software-implemented 'execution engine' resides on the user terminal (an MHEG-5 engine) that on-the-fly translates part of the passive content into active one.

## 5.4.2    The user presentation interface model

MHEG was invented with the aim of coding interactive multimedia presentation in a way capable of offering them to consumers having no prior PC experience. So, the platform's interface with the user had to take into account this characteristic. The MHEG-5 interface model was designed to allow the development of applications-presentations compatible with that requirement.

According to the adopted model, a presentation consists of a set of scenes one succeeding the other either on user's instructions, or at the occurrence of a synchronization event. Each scene contains visual and/or audible elements. For instance, in the case of a tourist information scene, the user would listen to a region's traditional song while at the same time viewing certain distinctive natural characteristics of that area.

Depending on the feedback the application accepts from the user, a new scene is selected to follow current one. Each scene contains all necessary interfacing elements giving the user control over its elements. For instance, the user can select a checklist or instruct transition to another scene. Execution of an MHEG presentation is characterized by the sequence through which individual scenes are presented to the used. This model of interaction is akin to the very popular navigational model of the *World Wide Web* (WWW). In the web, this model is implemented using the *Hyper-Text Markup Language* (HTML) notation, whereas MHEG scenes correspond to individual HTML pages. While the models of interfacing with the user of these two technologies are alike, their architecture model differs. More precisely, the MHEG model is based on objects and a machine independent engine for executing them, while HTML follows the more simplified approach of a relatively unstructured textual description of the web pages contents.

### 5.4.3   The network model

An MHEG application is a set of objects derived from the classes defined in the MHEG hierarchy. Subsets of these objects are grouped into objects of class `Scene`, which represent the wrapping elements of an actual scene presented to the user. Object of class `Scene` are therefore the elements that contain information on all objects that are displayed and interact with the user in the course of a scene. An entire presentation is therefore composed of a number of scenes containing in turn other objects. Scenes are typically serialized and stored in a server, which the user can access to view the presentation.

As explained during our description of the MHEG's user-interface model, scenes can change either depending on the feedback they get from the user or according to time scheduled events. Individual scenes (and the objects they contain) are therefore needed by the user's terminal only at the moment in which the appearance of the scene they belong is instructed. Figure 5.4 depicts an example of a network interactive multimedia presentation system based on MHEG.

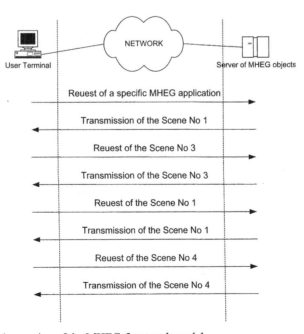

**Figure 5.4**  Typical operation of the MHEG-5 network model

In Figure 5.4, the transmission of objects constituting an MHEG application does not take place before the actual execution of the service starts, but occurs on demand in response to user's interaction with the service. This model is advantageous in terms of reduced load on the network connection. The trade-off is that it suffers from a larger latency in user's requests for a change of scene. The scene objects already transmitted may need to be re-transmitted if the represented scene is requested for a second time. This is, however, a characteristic of the individual machine. In an optimized implementation, and if the terminal's hardware resources can afford it, scenes may be cached for a certain time frame so that in subsequent requests the responsiveness of the system will be much higher. Caching scenes will also have a beneficial effect on the transport resources of the network. In addition to the user interface model, the network model is also similar to that of HTML. In fact, as related to the network part, both technologies have common requirements. In particular, a protocol is needed that would provide the following capabilities:

- Unique identities of the elements of a presentation (references). It should be possible for each element of a presentation to have its own, network-wide unique identifier. In this way, unambiguous dereferencing in a distributed environment will be possible.
- A structured name space. Since the elements of a presentation (be these objects in the case of MHEG-5 or pages in the case of HTML) are stored at various network nodes, an addressing mechanism is required that will identify the relevant network nodes. This presupposes a structured name space and a definition of its scope, along perhaps with a hierarchical structure in order to facilitate distribution and reduce name pollution.
- A directory service. This service will allow search and retrieval of the components of a presentation. This sort of facility is related, for instance, to the ability to issue a request from a server relative to a specific file that contains a given presentation element.

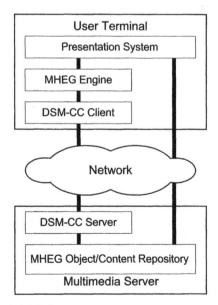

**Figure 5.5** DMS-CC's role in MHEG

In the case of HTML, that protocol exists and is the HTTP. In the MHEG case, the Digital Storage Media Command and Control (DSM-CC) (ISO, 1996) standard is used. Figure 5.5 depicts the role played by DSM-CC in providing MHEG-based multimedia services.

Let's examine more closely the use of DSM-CC by MHEG. When an MHEG application starts execution the service node from where object of the presentation will be requested is determined. So, DSM-CC sets a name space exclusively for the scope of the objects of that application. This service is called DSM-CC directory service. By using this facility, application objects, content files as well as stream content files can co-exist in a tree-like structure inside the multimedia server's file system. By limiting the scope in the file system that contains only the elements of the running application, names of individual objects need be unique only in the scope of the presentation. In this way, application element names are unambiguously defined. The following terms are used in the description of that facility:

- *Object Reference*: used to identify object deriving MHEG-5 classes.
- *Content Reference*: used to identify content media in an MHEG-5 presentation.
- *Group Identifier*: index to grouping objects (e.g. Scene).

### 5.4.4    The description model

An MHEG-5 application consists of the following modules:

- *Application Objects (AO)*: serialized forms of the objects stored in files. This is a term applied to the binary files that may contain any object derived from any class of the MHEG class hierarchy. There are also objects derived from the `Application` class, which are different.
- *Content Objects*: media content stored in files.
- *MHEG Engine*: the decoding and execution (interpreting) engine.

The MHEG Engine is a runtime environment running on the user's terminal. This environment can fetch objects and content of a specific MHEG application from the network and process them. The outcome of this processing is a multimedia presentation with which the user can interact at his terminal. This engine constitutes the 'framework' part of the application, which is used invariably for all MHEG-5 presentation. In contrast, objects and content files are authored or automatically produced with a specific application in mind. The MHEG-5 specification defines a class hierarchy through which MHEG Objects are produced.

Classes of the MHEG-5 hierarchy in addition to attributes define a certain dynamic behavior that specifies how they will react to external stimuli (e.g. interaction by the user). This dynamic behavior is not implemented in the passive objects produced by the classes, but it is part of the execution engine's implementation. Concluding, the MHEG-5 specification defines a set of classes comprising (in general) both dynamic behavior as well as data. The data are mainly stored as attributes in the passive objects derived from the class hierarchy, while the dynamic behavior is mainly implemented in the MHEG-5 engine, and is general for all presentations. Content files comprise visual and audible data that are opaque to the machine, but are useful to the human user (e.g. text, images, sound, video).

The MHEG-5 specification describes the following properties for the defined classes:

- *Attributes*: all attributes are either data or variables of objects derived from a class.
- *Inherited attributes*: these are data values inherited from ancestral classes (according to the object oriented inheritance model).
- *Own exchanged attributes*: these are variable that are not inherited, and are stored along with passive objects to be transmitted through the network.
- *Own internal attributes*: these attributes do not necessarily correspond to variables and characterize mainly the state in which an object of the class is found at each point during its execution.
- *Events*: each object derived from a class can produce events which affect other objects (probably causing them to produce more events, and so on).
- *Internal behaviors*: the behavior of the class' objects is described here in basic administrative functions (e.g. preparation and activation of an object).
- *Effect of MHEG-5 actions*: every object, during the course of an application's execution can accept external stimulation in the form of actions.
- *Formal description*: a formal notation is used here, the Backus – Naur Form (BNF) to describe the structure of the passive object that is produced by the class.

Of all the class properties that are described in the MHEG standard, only exchanged attributed are coded and transported within passive objects. The structure of these objects is described using *Abstract Syntax Notation 1* (ASN.1). The structure is then serialized for storage or transmission over the network according to the *Basic Encoding Rules* (BER) standard. When authoring a new MHEG presentation, therefore, values are assigned only to the own exchanged attributes of objects. The rest of the properties are implemented inside the MHEG engine.

## 5.4.5    The class hierarchy

The MHEG model is based on object oriented programming principles. Those principles are mostly evident in the process of creating new MHEG presentations, whereby everything boils down to extending, interrelating and parameterizing objects derived from a given class hierarchy. Design and implementation issues of MHEG engines are not described in the standard, and can in fact be carried out using structured programming techniques. This paragraph describes the MHEG-5 class hierarchy, and examines the behavior of its most important classes.

Figure 5.6 depicts the MHEG-5 class hierarchy diagram. The classes whose names are printed in boldface text are concrete classes (meaning that objects can derive from them), whereas the rest of the classes are abstract (no objects can be instantiated). The most fundamental classes of the hierarchy are detailed in the subsequent paragraphs.

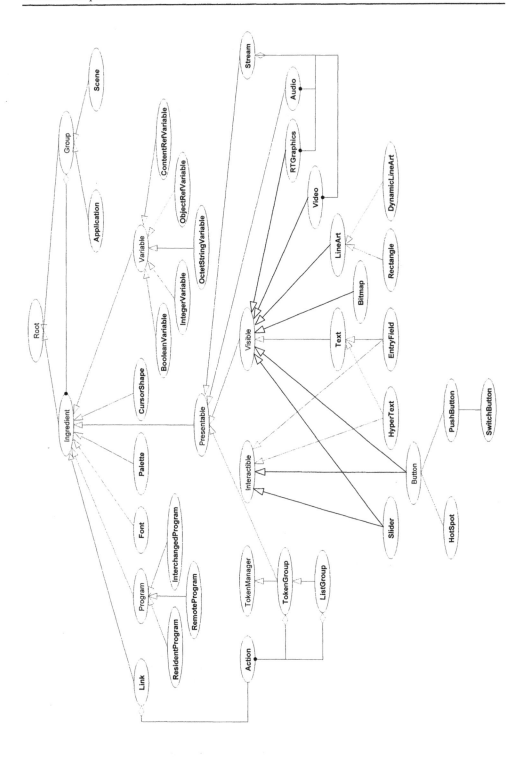

**Figure 5.6** MHEG-5 class hierarchy

### 5.4.5.1    Root

This is the base class of the entire hierarchy from which all other classes descend. Whatever is defined here is inherited and is present in all other classes. The root class defines:

**Own exchanged attributes**
- Object identifier. This is the unique identifier of each object within the name space each presentation belongs to. The structure of an object identifier is composite as it comprises the group identifier and the object number. The group identifier quite expectedly identifies the group to which the object belongs. The object number uniquely identifies the object within the said group.

**Own internal MHEG-5 attributes**
- Availability status: it indicates whether the passive part of the object is available to the MHEG machine for activation or not.
- Running status: it indicates whether the passive part of the object has started execution in the MHEG machine or not.

**Events**
- `IsAvailable` is emitted once the object becomes available
- `IsDeleted` is emitted once the object is destroyed
- `IsRunning` is emitted once object is activated
- `IsStopped` is emitted once object is de-activated.

**Internal behaviors**
- `Activation`. Activating an object means that a predetermined behavior is set in motion, and that this activation effects the condition of the object rendering it 'active'. At any given point in time, only a single object of the application can be active. Often, the effect of an object's activation survives its deactivation (e.g. in the case of a Bitmap whose activation causes it to be rendered in the screen).
- `Deactivation`. It corresponds to the revocation of an activation condition, and in certain cases the effects of activation are also revoked. It emits an `IsStopped` event.
- `Preparation`. It signals the object as being available. If the information of the passive object is not available at the terminal, it is requested and is transmitted to the terminal from the MHEG server. It emits an `IsAvailable` event.
- `Destruction`. It de-activates and in turn frees resources that had been committed to cater for its availability. It emits an `IsDeleted` event.

**Effect of MHEG-5 actions**
- `GetAvailabityStatus` returns the object's availability status.
- `GetRunningStatus` returns the object's activation status.

### 5.4.5.2    Group

This is an abstract class that encompasses all the characteristics that are used for the grouping of objects composing an MHEG-5 application. The objects this class wraps are of type `Ingredient` (see below). This is actually a use of polymorphism. Class `Ingredient` defines certain methods (e.g. activation/deactivation) that are actually implemented in the derived classes. The `Group` object invokes these methods on objects it sees as of type `Ingredient`, but actually the implementation of the methods defined in derived classes are called. The basic properties of class `Group` are:

**Own exchanged attributes**
- OnStartUp. It defines a set of elementary actions that must be executed by the MHEG engine at the end of the object's activation.
- OnCloseDown. It defines a set of elementary actions to be executed prior to calling Deactivation on that object.
- GroupCachePriority. An indicative value that can be put to use as a quantitative indication of the need for caching for each object fetched to be executed by the engine. This value can also be interpreted as the probability that the object will be requested more than once.
- Items. A set of objects of type Ingredient. These objects constitute the set of objects that are visible to the user during a Group object's activation period. Such objects are those derived from classes Application and Scene.

### 5.4.5.3    Application
This class is a descendant of class Group. By assuming the properties of the Group class, the Application class has the ability to group together objects of class Ingredient (or their descendants). There is a constraint involving the Application class. Namely, that one and only one object of this class can be active at any given point, and that it must be in an active case to allow objects of other classes to be activated.

Every MHEG presentation is composed of one object of class Application and one or more objects of class Scene. Both have a common ancestor in class Group, and their role is to aggregate objects of class Ingredient. So, an MHEG presentation is stored as a set of files each corresponding to a Scene or the one Application object. The Application object is in fact the first object that is requested and whose activation triggers the beginning of the presentation. The basic characteristics/behaviors of class Application are:

**Own exchanged attributes**
- OnRestart. A set of actions to be triggered when the Application object is restarted.

**Own internal attributes**
- DisplayStack. A sorted list of references to objects of class Visible which determines how they are organized in a visual form. Objects found at the end of this list are located on the background, while the object at the top of the list is on the foreground.

**Effect of MHEG-5 actions**
- Launch activates a new application-MHEG presentation after stopping the application that is currently active.
- Spawn suspends the currently active application and activates a new one, upon the finishing of which, the suspended one resumes.
- Quit quits the running application and resumes the suspended one (if any).
- LockScreen freezes the screen so as to forbid changes on objects of type Visible.
- UnlockScreen allows objects of type Visible to be modified again.
- OpenConnection sets up a connection through some protocol and assigns an index to that connection so that it can be referenced by other objects.
- CloseConnection closes an open connection.

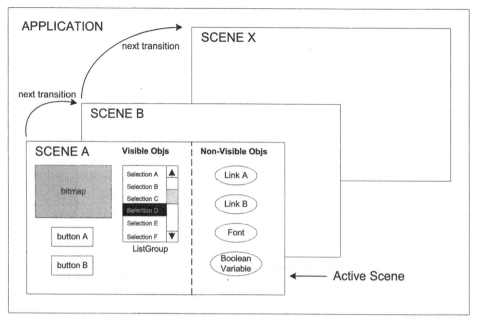

**Figure 5.7**  Grouping of objects into different scenes

### 5.4.5.4    Scene

This is a very important class. Objects of it group together objects of type `Ingredient`. Its purpose is to allow temporal and spatial synchronization of a presentation. Typically, more than one objects of class `Scene` are included in an Application, but only one of them is active at any given point. The basic characteristics-behaviors of this class are:

**Own exchanged attributes**
- `SceneCoordinateSystem`, screen's coordinate system dimensions.
- `AspectRatio`, screen dimension's ration (default vale 4/3).
- `NextScenes`, a list of indices to objects of class `Scene` (`Group Identifiers`) which may (depending on the user's interaction with the current scene) follow after the currently active one. This lists also indicates relative weights which offer a quantitative indication of the likelihood that a certain scene object will follow. This type of information can be used by the MHEG machine for prefetching `Scene` objects in order to improve system's responsiveness on user indications for a change of scene.

**Own internal attributes**
- `Timers`. This is a list of timers that denote time instances upon which events of type `Timer` will be generated.

**Events**
- `UserInput`. This event is generated when there is user input available. It also carries parameters indicating the type of user input.
- `TimerFired`. This event is generated when a certain timer expires.

**Effect of MHEG-5 actions**

- `TransitionTo (ConnectionTag, TransitionEffect)`. Causes a change in scenes. In particular, this action refers to the scene object that is to be displayed. As soon as this method is called, the current scene is withdrawn along with all objects of class `Ingredient` it may hold, and the new scene object is activated. The transition effect parameter specifies a visual effect to couple this transition between two scenes (e.g. fade away). Depending on the user's interaction with the presentation, different scene activation sequences may take place. Figure 5.7 depicts the grouping of visible and non-visible objects into scenes and that of scenes into a single Application Object. Figure 5.8 depicts different activation graphs in response to different user inputs.
- `SetTimer (TimerID, TimerValue)`. Sets specific values to a scene object's timers.
- `SendEvent( EventSource, EventType, EmulatedEventData)`. Generates the event that is indicated.

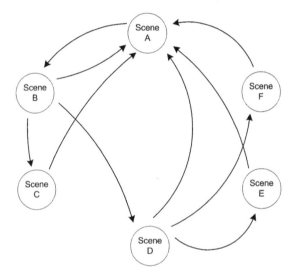

Two examples of possible sequences of Scene activation
according to the graph

A - B - D - F - A - C

A - B - C - A - B - D - E - A

**Figure 5.8** Different activation sequences depending on user input

## 5.4.6    How an MHEG-5 engine works

In Section 5.4.5 we examined the most fundamental classes of MHEG-5 class hierarchy in order to form a notion about what the MHEG objects consist of. We saw that an MHEG object deriving from an MHEG class has attributes, events, internal behaviors and effect of

MHEG-5 actions. Interpreting these properties in analogy with common object oriented terms, we can say that:

- The *attributes* are the data members (where the *own exchanged attributes* are their initial values).
- The *internal behaviors* and *effect of MHEG-5 actions* are the methods (where the internal behaviors represent fundamental functions).
- The *events* are primitive messaging elements that may be produced by that object and cause the invocation of the method of another one.

The last property, events, is illustrated only in the description of the classes of the MHEG hierarchy in order to provide information for the implementation of an MHEG engine. It does not constitute a concrete element in an MHEG object with the programming interpretation of the term. However, besides the implementation of the objects themselves, events must have a concrete programmatic analogue in the MHEG engine that may be for instance 'method invocation', or some other mechanism for communication among objects.

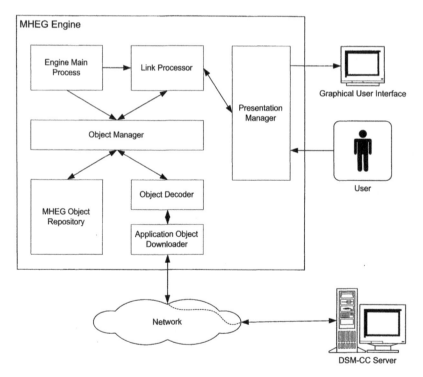

**Figure 5.9**  Internal architecture of a typical MHEG-5 engine

The only persistent part of an MHEG object that is stored on the application server is the initial values of its attributes, namely its own exchanged attributes. These values are defined using ASN.1 notation, and get their binary form according to BER. The binary form of the own exchanged attributes of the objects that constitute an MHEG-5 application is embedded in the Application Objects that this application consists of. In Figure 5.9 the internal architecture of an MHEG-5 engine at a coarse level is depicted.

When an MHEG-5 engine begins execution of that application, the first Application Object (the first one is always of type `Application`) is downloaded to the engine by a module called *Application Object Downloader*. Another module called *Object Decoder* parses the binary format of that Application Object and the module *Object Manager* instantiates all the MHEG objects that their *own exchanged attributes* are contained in the AO. Actually, an AO contains just one object of class either `Application` or `Scene` (see Section 5.4.5.2). However, such an object encompasses in its attributes multiple objects of class `Ingredient`. Therefore, the decoding process of an AO results in the instantiation of many objects of type `Ingredient`, that will be stored in the *MHEG Object Repository*, and the initialization of the objects' attributes with the values encoded in the AO.

The subsequent AOs following the first one are of class type `Scene`. At any time only one `Scene` object and its `Ingredient` objects are instantiated and activated in the MHEG Object Repository. The *Main Engine Process* invokes the `activation` method on them, and this may cause the generation of various events that are being forwarded to the *Link Processor*. Link Processor has previously instantiated all the `Ingredient` objects that are of type `Link`. Whenever Link Processor receives an event generated by an `Ingredient` object, it checks if this event *fires* any active `Link` object. If it does, the Link Processor executes the MHEG-5 Actions that this `Link` object contains which in turn result in the invocation of the corresponding methods (Effect of MHEG-5 Actions) of the active `Ingredient` objects that are targeted to. Then the objects that have a GUI representation invoke methods directly to the *Presentation Manager* that is eventually responsible for the execution of the presentation.

If we look behind the specialization of the MHEG standard for multimedia presentations, we can see that the programming model that introduces is reusable in the domain of telecommunication networks. As we have examined how an MHEG-5 engine works, we can point out the following properties of the system as a whole:

- The *programmable system*, which consists of a combination of the MHEG Engine and the Application Objects, controls *resources* that are the GUI elements (buttons, videos, etc.) that interface directly with the user.
- The *resources* are organized into a finite number of *timely bound groups* – the `Scene` objects – that only one is active at a time.
- Each controlled *resource* is mapped in a *system object* – MHEG object.
- Furthermore, there are special purpose *system objects* like `Scene`, `Link`, `Action`, `TokenManager`, that enable the rest to interrelate to each other in a way that is programmable.
- Each *system object* has a *passive part* – its own exchanged attributes – and an *active part* – its methods.
- The *passive parts* are programmable and compose a *system's application* – an MHEG-5 application – and the *active parts* are programmed once and constitute an integral part of the *system's execution environment* – the MHEG-5 engine.

From the points stressed above, we see how the terminology of the MHEG system is mapped to more general terms that are often used in the domain of telecommunication networks. One can easily extend this analogy to figure out how another class hierarchy can be defined to create a system, based on the same design model, that for instance controls the services in an *Intelligent Network* (IN) environment (see Chapter 7). This atypical object oriented design model is an alternative approach, compared to more widespread object oriented software technologies like Distributed Objects (see Chapter 4) and Mobile Agents (see Chapter 6), to introduce programmability in telecommunication networks. In this book we will no longer elaborate on this 'MHEG approach'. Nevertheless, we presented it in this chapter to give the reader a more in-depth view of how alternative methods exist for introducing object oriented software practices in telecommunications.

# References

Bitzer H. and Höfrichter K. (1998) MHEG-5 – The Standard API for Digital Television, IBC Conference, 12 September 1998.

Eckel B. (1998) *Thinking in Java*. Prentice Hall. Available from www.eckelobjects.com, ISBN:0-13-659723-8.

ISO/IEC 13818-6 (1996) MPEG-2 Digital Storage Media Command and Control.

ISO/IEC 13522-5:199 (1998) MHEG-5, Support for base level interactive applications, June.

Krupczak B. and Calvert K. (1998) Implementing Telecommunication Protocols in Java. *IEEE Communications Magazine*, **38**(10).

Lindholm T. and Yellin F. (1996) The Java Virtual Machine Specification. Available from www.javasoft.com

Meyer T. and Effelsberg W. (1995) MHEG Explained. *IEEE Multimedia Magazine*, Spring issue.

Stroutstrap B. (1996) *The C++ Programming Language, 3rd ed.* Addison-Wesley.

# 6

# Agents

The term 'agent' has become a buzzword in the nineties and denotes a new software paradigm and related technologies, which are considered as the next step beyond object orientation and *Distributed Object Technologies* (DOT). This chapter is devoted to the concept of (software) agents. Therefore it describes the basic principles of agents and introduces the related concepts and terminology. In this regard, we look at the definition of agents in general, addressing both *Intelligent Agents* (IA) and *Mobile Agents* (MA), and look at available agent standards. Subsequently, we focus on Mobile Agents, and describe the basic capabilities of Mobile Agent systems. This is in line with Nwana (1996), who mentions that 'public perception of agent technology is synonymous with mobile agents'! In fact, we believe that this category of agents is more interesting for the telecommunications. Finally, we briefly sketch the application of MA within the telecommunications environment.

## 6.1 General Principles of Software Agents

In this section we look at the origins and definitions of software agent technologies, covering both Intelligent and Mobile agents.

### 6.1.1 Towards agents – how it all begun

In retrospect, the whole (successful) story of agents started in the early nineties with the parallel appearance of different agent concepts and technologies, which we today would separate into Intelligent Agents and Mobile Agents (Guilfoyle, 1994; CACM, 1994). In fact, the term agent was used as a buzzword in these days, since not everything labeled as an agent was based on agent concepts or agent technologies as we understand them today. For example, the term 'agent' has also been used for many years in the field of distributed computing, where it refers to specific (client or server) entities used in solving specific tasks in a distributed computing system (e.g. directory system agents, mail user agents, management agents, etc.).

The interest in IA was coined by the increasing notion of *Multi Agent Systems* (MAS) and Interface Agents in the early nineties, driven by the *Distributed Artificial Intelligence*

*Object Oriented Software Technologies in Telecommunications,* Edited by I. Venieris, F. Zizza and T. Magedanz
© 2000 John Wiley & Sons Ltd

(DAI) research community (Wooldridge, 1995). The MAS concept is largely conceived upon the idea that complex activities are the outcome of interactions between relatively independent software entities called 'agents'. A MAS may be defined as a set of agents that interact with each other and with the environment to solve a particular problem in a co-ordinated (i.e. behaviorally coherent) manner. Therefore, agent communication and cooperation represents a major issue of this type of agents.

Interface agent research has considered the coupling of automated software controlled systems to human reasoning ability, thus completing the chain of being able to provide human invoked automated software control systems. The pioneering work of Pattie Maes' team at the MIT has defined many of the factors that need to be resolved (Maes, 1994). Usually this type of agent has acted as Advisory Agents (e.g. intelligent help systems) or Personal Assistants that are supporting human users during their daily work. The major goal of these agents is to collaborate with the user, and hence the main emphasis of investigations clearly lies in the field of user/agent interaction. Some of these agents featuring an interface to the network, have become famous as 'smart mailboxes' and 'search engines'. Not only do they provide an intelligent interface to the user, but they also make extensive use of the various services available in the network. In contrast to smart mailboxes which perform advanced mail filtering based on the user's preferences, search engines collect knowledge available in the network on the user's behalf. This type of agent has also been called 'KnowBots' or 'Softbots' (Etzioni, 1994).

Nevertheless, probably the most important boost in creating awareness for the term 'agent' was the appearance of the Mobile Agent concept, coined mainly by a new technology called TeleScript developed by General Magic in 1994 (White, 1994). It was the time where scripting languages such as the Tool Command Language (Tcl) and its derivative SafeTcl (Borenstein, 1994) gained much attention, since they enabled rapid prototyping and the generation of portable code. The concept of smart messaging (Reinhardt, 1994), based on the encapsulation of SafeTcl scripts within emails (Borenstein, 1992), made new mail-enabled applications possible. The concept of mobile computing has gained increasing importance, enabled through *Mobile Agent Technology* (MAT) (Chess, 1995). Furthermore, the telecommunications domain has been considered as a main application area for Mobile Agents (Magedanz, 1996b). Last but not least, it was the beginning of the Java-age, and Java is today the basis for most mobile agent systems.

But let's consider TeleScript a little bit deeper. TeleScript (White, 1994) introduced as the 'PostScript language for the network', was more than just a language, as it provided a complete mobile *Agent Execution Environment* (AEE). This environment was designed to enable the implementation of the concept of remote programming, which was proposed as an alternative approach to the good old *Remote Procedure Call* (RPC).[1] The main idea was to ship the (small piece of) code to the data and not the (large amount of) data to the code. Whereas at this time Safe-Tcl and Java were primarily used to enable asynchronous operation and remote execution of 'mail-enabled' and 'WWW-enabled' applications within the Internet, the metaphor used within Telescript was the 'electronic market place'. Within this market, agents asynchronously perform tasks on behalf of users. They may communicate with the user, with the services available in the network or with other agents. In order to perform specific tasks, the agents migrate through a network to visit remote

---

[1] The PostScript language can be considered a rudimentary form of the more general idea of sending programs to and executing them at a remote site as it involves sending print programs to a remote processor in a printer.

sites. This means that during its execution, a MA may move from node to node in order to progressively accomplish its task. In other words, agents are capable of suspending their execution, transporting themselves (i.e. program code, data, and execution state) to another node and resuming execution from the point at which they were suspended. However, due to its closed architecture and the coincident increasing acceptance of Java as the universal scripting language, and thus the adoption of Java as programming language for all state of the art MA systems available today, TeleScript was abandoned in 1998 and has been replaced by a Java-based agent platform, called Odyssey, with less impact and success. Nevertheless, a lot of the concepts originating from TeleScript are still present in existing Mobile Agent systems and standards.

Today there has been a lot of development and general excitement in the area of MAT, much of which has evolved from the platform independence of the Java language with its built-in object serialization and network communications support.

We can summarize that based on this coincidence of the appearance of various agent concepts, agents became a fashion technology for the research and development community. However, this created also confusion, since there was a lack of common definitions and standards, resulting in various concepts, languages, architectures, technologies and terminologies. However, this situation has changed a little with the establishment of common agent standards.

Before we look in more detail at available and emerging agent standards, we want to introduce the two major agent categories considered today, namely Intelligent Agents and Mobile Agents, in some more detail.

## 6.1.2   Agent definitions

The term 'Software Agent' (Bradshaw, 1997) has been adopted as the most general phrase to describe the concept of an autonomous software entity that automates some of the tasks of a human or another software process which have been delegated to it. In other words, an agent is an encapsulated software entity with its own state, behavior, thread of control, and an ability to interact and communicate with other entities – including people, other agents, and legacy systems.[2] This definition puts an agent in the same family, but distinct[3] from, objects, functions, processes, and daemons. The agent paradigm is different to the traditional client-server approach; agents can interact on a peer-to-peer level, mediate, collaborate, and cooperate to achieve their goals.

However, there is no unique definition of an (Software) Agent. This reflects the diversity of theories, languages, architectures, technologies and standards. Nevertheless, today we can simplify our considerations on agents by reducing the whole spectrum of available concepts to the two main categories, namely Mobile Agents and Intelligent Agents. Mobile Agents embody the ability to migrate seamlessly from one platform to another whilst retaining state information, typically with limited intelligence. By contrast, an Intelligent Agent is an agent that exhibits 'smart' behavior. Such 'smarts' can range from the primitive behavior achieved through following user-defined scripts, to the adaptive behavior of neural networks or other heuristic techniques. In general, Intelligent

---

[2] Not necessarily all of these for every instance of an agent.

[3] An agent is at a higher level of abstraction.

Agents are not mobile since the larger an agent is, the less desirable it is to move it; coding artificial intelligence into an agent will undoubtedly make it bigger.

Until recently, there has been a distinct line drawn between these paradigms and the two research domains have been focusing on quite different problems. However, the two research areas have been relying on a similar conceptualization, that of utilizing separate software processes (they may be implemented as procedure calls, a thread or several threads, but look and behave conceptually as autonomous processes) for dealing with automating control tasks and it is also evident that the more specific attributes of each are merging towards a unified approach. This means that we are heading towards Mobile Intelligent Agent platforms which enable the designer to accommodate agents with both characteristics.

Nevertheless, we want to present in the following the main characteristics of both agent types in order to clearly identify their major contribution to the interest in agents.

### 6.1.2.1    Intelligent agents

Probably the most important attribute that distinguishes this type of software agents from other types of software processes is their ability to cooperate in some manner with other agents (human or software), either to gain a local advantage themselves (exploiting the actions of other agents – e.g. by being competitive) or to add to the global 'good' of the system in which they operate (i.e. via collaboration). Therefore, they are sometimes also called Intelligent Co-operative Agents. The abilities of software agents have been described eloquently by Wooldridge (1995) and can be classified into the possession of Social Ability, Autonomy, Reactivity, and Adaptability. Wooldridge (1995) provides a diverse review of intelligent software agent research the interested reader is directed to. We will just concentrate in the following on the aspects of agent communication and cooperation, as these are central for this type of agent.[4]

Communication enables agents in a multi-agent environment to exchange information (e.g. to allow an agent to inform another agents about its current beliefs, amount of resources available, additional information about its environment, etc.) on the basis of which they originate their action sequences and cooperate with each other. Software agents generally communicate with other agents in order to work more flexibly. To achieve coordination, the agents have to interact and exchange information and perhaps to negotiate. This is done by means of an *Agent Communication Language* (ACL), which is a language with precisely defined syntax, semantics and pragmatics. However, it has to be stressed that agent communication is accomplished through the use of three components: the ACL, content language, and ontology, this is a common approach for agent systems (see also Figure 6.1). Ontology enumerates the terms comprising the application domain and resembles a data dictionary in a traditional information system. The content language is used to combine terms in the ontology into sentences (logical or otherwise) which are meaningful to agents who have committed to this ontology. Sometimes the ontology and content language are so tightly integrated that they become the same thing (i.e. a list of sentences is the content language) which represent the ontology. Finally the ACL acts as a

---

[4] We do not intend to provide a comprehensive definition and introduction to intelligent agent technology and the related AI, DAI background as this would be beyond the scope of this chapter and this book. Therefore, we refer the interested readers to Wooldridge (1995) for aspects of Intelligent Agent theory, languages and architectures and Nwana (1996) for a detailed discussion on background and the application domains of software agents.

protocol, enabling the development of dialogues containing sentences of the content language between agents and defining certain semantics for the behavior of agents participating in such dialogues.

Most ACLs derive their inspiration from speech act theory (Searle, 1969), which was developed by linguists in an attempt to understand how humans use language in everyday situations, to achieve everyday tasks, such as requests, orders, promises, etc. It is based on the idea that with language the speaker not only makes statements, but also performs actions. A speech act can be put in a stylized form that begins 'I hereby request ...' or 'I hereby declare ...'. In this form the verb is called the performative, since saying it makes it so.

**Figure 6.1** Intelligent agents

The probably most prominent ACL, representing the *de facto* standard before *Foundation for Intelligent Physical Agents*'s (FIPA) ACL (FIPA, 1998) became available (see the next section), was the *Knowledge Query and Manipulation Language* (KQML) (Finin, 1994), which defines a framework for knowledge exchange. KQML focuses on an extensible set of performatives, or message types, which define the permissible operations that agents may execute on each other's knowledge and goal stores. As the content of the messages was not part of the standard, the *Knowledge Interchange Format* (KIF), a formal language was defined based on first-order predicate calculus for interchanging knowledge among disparate computer programs. KQML and KIF have been developed in the context of the DARPA Knowledge Sharing Effort (Ginsberg, 1991).

Some of the benefits cited in the literature for intelligent agents, i.e. MAS, can be summarized as:

- To address problems that are too large for a centralized single entity, for example because of resource limitations or of robustness concerns (the ability to recover from fault conditions or unexpected events).
- To enable the reduction of processing costs. It is less expensive (in hardware terms) to use a large number of inexpensive processors than a single processor having equivalent processing power.
- To improve scalability and adaptability. The organizational structure of the agents can dynamically change to reflect the dynamic environment, i.e. as the network grows in size the agent organization can re-structure by agents altering their roles, beliefs and actions that they perform.
- To provide solutions to inherently distributed problems, i.e. where the expertise is distributed.

At a high level the multi-agent systems approach is intuitively simple. Developers can draw on their experience in solving problems in real world co-operation with others. Also, the MAS paradigm is inherently scalable due to modularity, loose coupling between interacting elements, and higher levels of design abstraction (the level of abstraction is greater than that of the object level).

However, the other side of the coin reveals that working with MAS requires a solid background in artificial intelligence and DAI, which is not commonly available. Furthermore, this approach of decentralization, based on distributed autonomous entities communicating via quite complex ACLs, departs from existing software solutions. In fact, the success stories of DAI solutions on a global scale are quite limited. This aspect makes the integration of legacy technologies into MAS quite complex, and hinders a smooth migration towards MAS solutions.

### 6.1.2.2    Mobile agents

Mobile Agents, also referred to as transportable agents or itinerant agents, are based on the principle of code mobility. Mobile code enhances the traditional client/server paradigm[5] by performing changes along two orthogonal axes:

1.  Where is the know-how of the service located?
2.  Who provides the computational resources?

Three main paradigms for mobile computations have been identified (Fugetta, 1998). These are: Remote Evaluation, Code On Demand, and Mobile Agents. These paradigms differ in how the know-how, the processor and the resources are distributed among the components of a distributed system. The know-how represents the code necessary to accomplish the computation. The resources (i.e. data) are that located at the machine that will execute the specific computation.

-   In the Remote Evaluation (REV) paradigm (Stamos, 1990) a component A sends instructions specifying how to perform a service to a component B. The instructions can, for instance, be expressed in Java's bytecode format. B then executes the request using its resources. Java Servlets are an example of remote evaluation.
-   In the Code on Demand (CoD) paradigm the same interactions take place as in REV. The difference is that component A has the resources collocated with itself but lacks the knowledge of how to access and process these resources. It gets this information from component B. As soon as A has the necessary know-how, it can start executing. Java Applets come under this paradigm.
-   The Mobile Agent paradigm is an extension of the REV paradigm (White, 1994, 1997). Whereas the latter focuses primarily on the transfer of code, the mobile agent paradigm involves the mobility of an entire computational entity, along with its code, the state, and potentially the resources required solving the task.

---

[5] In the client-server paradigm the server is defined as a computational entity that provides some services. The client requests the execution of these services by interacting with the server. After the service is executed the result is delivered back to the client. The server therefore provides the knowledge of how to handle the request as well as the necessary resources.

In other words, adopting the MA paradigm, component A has the know-how capabilities and a processor, but it lacks the resources. The computation associated with the interaction takes place on component B that has a processor and the required resources. For example, a client owns the code to perform a service, but does not own the resources necessary to provide the service. Therefore, the client delegates the know-how to the server where the know-how will gain access to the required resources and the service will be provided. An entity encompassing the know-how is a Mobile Agent. It has the ability to migrate autonomously to a different computing node where the required resources are available. Furthermore, it is capable of resuming its execution seamlessly, because it preserves its execution state.

This means that a Mobile Agent is not bound to the system where it begins execution. It has the unique ability to transport itself from one system in a network to another (see also Figure 6.2). The ability to travel permits a MA to move to a destination agent system that contains an object with which the agent wants to interact. Moreover, the agent may utilize the services of the destination agent system. When an agent travels, its state and code are transported with it. In this context, the agent state can be either its execution state, or the agent attribute values that determine what to do when execution is resumed at the destination agent system. The agent attribute values include the agent system state associated with the agent (e.g. time to live).

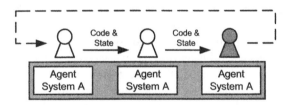

**Figure 6.2**  Mobile agents

The MA paradigm is important for network-centric systems because it represents an alternative, or at least a complementary, solution to traditional client-server architectures (Chess, 1995). Such solutions may contribute to a reduction of the overall communication traffic in network. For example, mobile code has the ability to engage with a server locally for searching large databases. The proximity to the server ensures the availability of high bandwidth for the communication.

Mobile agents are typically developed by means of machine-independent programming languages, i.e. scripting languages, where initial pioneering languages such as Safe-Tcl and Telescript are today mostly replaced by Java, due to its inherent portability and platform support. Nevertheless, it has to be stressed that the native capabilities of Java are not yet sufficient for implementing right away mobile agents. Extra functionality has to be implemented for realizing a MAT to support agent transport, mobility management, and security (see Section 6.3).

Looking at Mobile Agents, the following benefits have been most often cited (see in particular Harrison (1995) and Chess (1998)):

- Asynchronous/autonomous task execution. After the injection of an agent into the network environment the user can perform other tasks.
- Reduction of network traffic and of the necessary client processing power. Since massive data exchanges are handled locally at the nodes hosting the data, client computers could concentrate on performing only limited local tasks.
- Increased robustness. Reduction of dependence of network availability and client/server availability. Once the agent arrived at a target system the client may crash or the network may become unavailable without any drawbacks on the task processing.
- Automation of distributed task processing. Agents have itineraries that determine what tasks they have to perform where without requiring any user interaction.
- Decentralized/local task processing (control & management). Agent cloning enables the (automated) distribution of formerly centralized programs.
- Flexibility. On-demand software distribution/service provisioning – service software within mobile agents can be instantly downloaded to client and server nodes.

This means that Mobile Agents provide flexibility in (re-)distributing intelligence inside a distributed network environment, particularly for reducing network load and optimizing service performance. Due to its benefits listed above, various problems and inefficiencies of today's client/server architectures can be handled by means of this new paradigm.

The negative side of this technology is the security risks introduced, as a computer virus represents also some kind of MA. Furthermore, an agent may be attracted, modified or deleted by a hostile agent platform. Another typically stated and obvious concern is the question, if agent migration is always advantageous in comparison with message passing. For example, it is probably better to interact for small information exchanges by message passing in case the agent code is bigger than the expected data volume to be exchanged.

In summary, it has to be stated, that agent technologies have a lot of appealing advantages compared to traditional technologies for solving specific environmental requirements. But they imply the introduction of new (agent) platforms enabling mobility and/or advanced inter-agent communication in the target environment and require the adaptation of existing interfaces to that new agent environment. Adopting a more general view, legacy technologies, including Distributed Object Technologies, have also advantages in specific environments and more importantly a big installed base. For several applications RPCs still represent a powerful and efficient solution. Thus, an integrated approach is desirable, combining the benefits of both client/server and agent technology, and on the other hand eliminating or at least minimizing the problems that rise if one of these techniques is used as 'stand-alone' solution. Therefore, an integration of agent technologies with existing technologies represents the best solution to combine their advantages. This is of particular importance for agent standards.

## 6.2    Agent Standards

The international standardization of agents started relatively late in 1996/97. In principle, two main fora have to be considered in the context of agent standardization today, namely FIPA focusing as the name suggests mainly on Intelligent (co-operative) Agents, and the *Object Management Group* (OMG), which has initially investigated Mobile Agent system interoperability aspects and is today looking within a dedicated Agent *Platform Special*

*Interest Group* (PSIG) at the integration of DOT and agent technologies, comprising both Intelligent Agents and Mobile Agents.[6] In the following, we look in more detail at FIPA and OMG and their related agent standards.

## 6.2.1   Foundation for intelligent physical agents

FIPA is a non-profit association registered in Geneva, Switzerland.[7] FIPA's purpose is to promote agent technology through the development of specifications that maximize interoperability across agent-based applications, services and equipment. FIPA specifies the interfaces of the different components in the environment with which an agent can interact, i.e. humans, other agents, non-agent software and the physical world. The main emphasis in FIPA is on standardizing agent communication, i.e. a dedicated ACL is used for all communication between FIPA Agents. It has to be noted that also the agent platform is modeled in terms of specific agents with dedicated ACL interfaces (see below). In general, FIPA produces two kinds of specifications:

- Normative specifications mandating the external behavior of an agent and ensuring interoperability with other FIPA-specified subsystems.
- Informative specifications of applications providing guidance to industry on the use of FIPA technologies.

FIPA proposes the concept of an *Agent Platform* (AP) offering three basic services. These services are namely the *Agent Management System* (AMS), the *Directory Facilitator* (DF) and the *Agent Communication Channel* (ACC). Agents are considered residing on a *Home Agent Platform* (HAP) if they are registered on the home agent platform's AMS. Agents may offer their services to other agents and make their services searchable in a yellow pages manner by the Directory Facilitator if they register on the Directory Facilitator. Registration on a Directory Facilitator is optional while registering on the AMS is mandatory. Finally, the Agent Communication Channel is enabling Agent Communication (AC) between agents on a platform and between platforms by offering a message forwarding service called `forward`. Reachability between platforms is gained by making the forward service available by the OMG *Internet Inter-Orb Protocol* (IIOP).

The AMS is the core of any agent platform. It is responsible for registering agents on their HAP. An agent is residing on an AP that is then its HAP, if and only if it is registered with the agent platform's AMS. Registering on an AMS is done by calling the AMS' `message` method with a request to register encoded in an ACL string conforming to the FIPA ACL definition. Alternatively, registration can be done by calling the request method of the AMS with a data structure equivalent to an ACL message containing such data as sender, receiver and content of the request (the content in this case containing the agent

---

[6] It has to be stated that there was originally a third standards forum, namely the Agent Society, as a forum for exchanging agent information. Alignments with OMG and FIPA had been foreseen. However, the forum terminated its work in 1997. Nevertheless, there are rumors that the Agent Society will be reactivated in 1999. URL: http://www.agent.org

[7] FIPA homepage: http://www.fipa.org

description data). Other functionality offered by the AMS is deregistering of agents, modification of the agent description and modifying the agents lifecycle state.

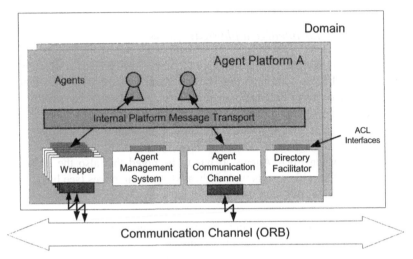

**Figure 6.3** FIPA agent management reference model

The DF is offering services similar to those of the AMS, but additionally offers search functionality. Thus the DF acts as a yellow pages directory, where agents willing to offer their services in a dynamic manner to other agents may register to a DF. The registration is done in the same way as with the AMS. Agents can deregister from the DF by calling the deregister service.

Basic to FIPA is the concept of Agent Communication by means of ACL defined in Part 2 of the FIPA specifications (FIPA, 1998). Agents have predictable behavior by common semantics defined in common interpretation of a common language. This is achieved by the concept of communication acts. The registering of an agent with an AMS is realized as a communication act of the action registration. In this communication act the roles of the agent and the AMS are clearly defined and the reactions of each party are determined by the state of the agent platform. If, for instance, the agent is already registered, it is clearly defined that it can not be registered again and an answer message denoting exactly this has to be send to the agent.

FIPA has chosen the concept of a communication channel per platform that is responsible for forwarding the ACL messages between agents. This is done via calling the forward service of the ACC. This is the only service or method being implemented by the ACC. Thus, the ACC is just offering the forward services, waiting for incoming messages to be forwarded to other agents. As platform local communication is free to the implementor of a platform, it is clear that the simplest solution for local communication between agents is realized by the platform native communication protocol. Inter-platform communication, that means communication between agents on different and possibly heterogeneous platforms, is mandatory to be realized by offering the forward service over IIOP. The most convenient method to do this is to use a common ORB implementation compliant to *Common Object Request Broker Architecture* (CORBA) 2.2 for offering that service.

In summary, it can be recognized that a FIPA Agent Platform provides the physical infrastructure in which (Intelligent) Agents can be deployed. An agent must be registered on (reside in) an agent platform to interact with other agents on that agent platform or other agent platforms. In fact, the concept of agent platform can be regarded as a refinement of the *facilitator* concept in the traditional intelligent agent frameworks.

Besides the AP, FIPA has also defined the logical concept of *domains*. An agent domain is a logical grouping of agents and their services, defined by their membership (registration) in a DF. Each domain has one and only one DF, which provides a unified, complete and coherent description of the domain. The DF lists all intelligent agents in the domain and advertises the intelligent agents existence, services, capabilities, protocols, etc. An agent may be present in one or more domains via registration in one or more DFs. A domain in this context can have organizational, geo-political, contractual, ontological, affiliation or physical significance.

## 6.2.2   Agent standardization within the Object Management Group

The OMG represents the global standards body for defining interoperable interfaces based on DOT. With its specification of the Common Object Request Broker Architecture, which represents the core of its *Object Management Architecture* (OMA), identifying different levels of services reusability, namely basic CORBA object services, common object facilities and dedicated application oriented frameworks, the OMG has established a globally accepted middleware standard for distributed systems. Various service specifications have been developed, providing standardized, implementation-independent interfaces and a common protocol, enabling a high degree of interoperability between applications of distinguished manufacturers. As AP can be considered a new middleware layer enabling agent-oriented programming, the relationships with OMA and particularly CORBA need to be studied. In this context, two main activities are worth mentioning; the early work of the *ORB and Object Services* (ORBOS) Group in between 1996–97 with the *Mobile Agent System Interoperability Facility* (MASIF) specification, and the *Agent Special Interest Group* (SIG) founded in the end of 1998 studying more comprehensively the relationships of agents and CORBA. The following Sections provide more details on these activities.

### 6.2.2.1   Mobile agent system interoperability facility

Recognizing the emergence of different MA systems, which are based on different approaches regarding implementation languages, protocols, platform architectures and functionality, the OMG aimed for a standard assuring interoperability between different *Mobile Agent Platforms* (MAP), and the (re-)usability of (legacy) CORBA services by agent-based components. Therefore, OMG ORBOS Group issued a Request for Proposal (RFP) for a MA facility in November 1995 (OMG RFP 3, 1995), which resulted in a corresponding MASIF specification adopted in 1997 (OMG MASIF, 1997). The standard comprises, among others, CORBA IDL specifications supporting agent transport and management, including localization capabilities.[8]

---

[8] It has to be stated that the target of agent transport between different agent systems is not fully enabled through the given specifications! Thus this transport capability would only become possible through mutual agreements on a common agent exchange format of agent system vendors.

The idea behind the MASIF standard is to achieve a certain degree of interoperability between MAPs of different manufacturers *without* enforcing radical platform modifications. MASIF is not intended to build the basis for any *new* MAP. Instead, the provided specifications shall be used as an 'add-on' to already existing systems.

As shown in Figure 6.4, MASIF has adopted the concepts of places and regions that are used by various existing agent platforms, such as Telescript. A *place* groups the functionality within an agency, encapsulating certain capabilities and restrictions for visiting agents. A *region* facilitates the platform management by specifying sets of agencies that belong to a single authority.

Two interfaces are specified by the MASIF standard:

- The MAFAgentSystem interface provides operations for the management and transfer of agents, whereas the
- MAFFinder interface supports the localization of agents, agent systems, and places in the scope of a region or the whole environment, respectively.

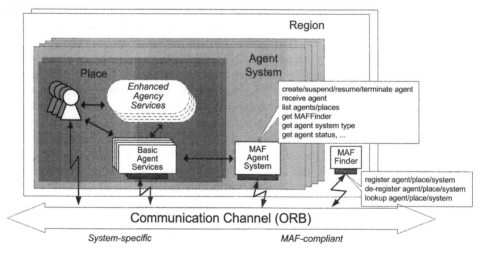

**Figure 6.4** OMG MASIF interfaces

As part of a MASIF-compliant agency, a MAFAgentSystem object interacts internally with agency-specific services and provides the associated CORBA interface to external users. In this way, it is possible to communicate with an agency either in an MASIF-compliant way (using the MAFAgentSystem interface) or in a platform-specific way (using platform-specific interfaces that may provide additional functionality, not handled by MASIF).

Apart from the agent specific CORBA interfaces MAFAgentSystem and MAFFinder, the MASIF standard explains in detail how existing CORBA services, like Naming, Lifecycle, Externalization, and Security Service, can be used by agent-based components to enhance the provided functionality. Note that the current MASIF specification only represents the first approach for a MA standard. It is believed that the work of the OMG Agent SIG will result in further specifications. For more details on MASIF see also OMG RFP 3 (1995) and OMG MASIF (1997).

### 6.2.2.2    Agent platform special interest group

The Agent Working Group of the Object Management Group, which is a joint activity of the Electronic Commerce *Domain Task Force* (DTF) and Internet Platform Special Interest Group, has been founded in late 1998.[9] The mission of the Agent Working Group as defined in its Mission Statement is to extend the OMA to support agent technology via the adoption of new OMG specifications, and recommend extensions to existing and emerging OMG specifications as appropriate. The main objectives are to:

- Recommend technology for adoption of common semantics, metamodel, and abstract syntax for agent technology, representing reusable, interoperable, portable application components.
- Enable developers to better understand how to develop applications using agent technology, thereby growing the market.

The scope of this work is on lifecycle services, declarative specifications, communication and conversation semantics, agent security, co-ordination and collaboration strategies, expression of belief, desire and knowledge, interoperability of agent technology, agent modeling and specification, testing, debugging, validation, and simulation of agent technology.

The Working Group aims to achieve its mission through the following activities:

- Collect information via  Requests for Information (RFIs).
- Develop an organized overview, survey, and possibly a reference architecture describing areas of agent technology and their relationship to each other.
- Develop a corresponding roadmap describing future technology adoption priorities.
- Initiate technology adoption processes via Requests for Proposals working with appropriate OMG Task Forces.
- Develop and distribute guidance material supporting the usage and application of agent abstractions and behavioral semantics in advanced distributed object systems.
- Dissemination of relevant information to OMG members.
- Liaison with other OMG Working Groups, Task Forces and Special Interest Groups.
- Liaison with external organizations and forums (e.g. FIPA).
- Informational presentations to educate OMG about agent technology.

At the time of writing, the Agent WG work has centered around:

- Production of an Agent Technology Green Paper Draft.
- Release of an Agent Technology RFI to industry in order to finalize the above Green Paper.
- Establishment of an OMG-FIPA Liaison.

It is likely, that the responses to the RFI will lead to the definition of new CORBA services, such as an Object Migration service, as proposed in Choy (1999).

---

[9] OMG agent WG home page http://www.objs.com/isig/agents.html

## 6.2.3    Integration of OMG MASIF and FIPA concepts

As it can be observed from the above presentations, OMG MASIF architecture puts more emphasis on the physical migration/operation support for Mobile Agents and on their physical object view, while FIPA emphasizes more on the logical and service view. This difference reflects the different focal points and constitutions of the two organizations.

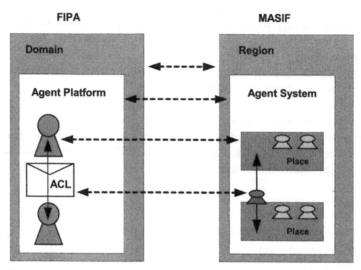

**Figure 6.5**   OMG MASIF vs. FIPA agent reference models

On the other hand, both the OMG MASIF Mobile Agent and FIPA Intelligent Agent approaches aim at adaptive and flexible co-operation and interoperability among dynamic autonomous systems. With the similarities between MA and IA technology that were discussed above, we can easily derive some correspondences between concepts in the MASIF and FIPA frameworks. Such correspondences will play an important role in the future evolutions and in the convergence of the agent standardization efforts. In this context (as illustrated in Figure 6.5):

- A *Mobile Agent* in the MASIF framework specifies a kind of message that migrates between software systems, and has the similar functionality as an Intelligent Agent *communication message* in FIPA framework. Current MASIF framework does not support MA communications. Such communication capability is, however, an integrated part of a MAP. With such communication capability, a MA will also have the features of an IA in FIPA.
- A *place* in the MASIF framework provides the operation environment for the Mobile Agents, with its capabilities (local services) and restrictions. Examples of such capabilities and restrictions can be database services, network resource management functions, security monitoring and management. All these capabilities are supported by special *Intelligent Agents* within the FIPA agent platform.
- The MASIF *Agent System* is the group of places that provides the platform for Mobile Agent migrations and operations. Therefore, a MASIF Agent System corresponds to a

FIPA *Agent Platform*, and partially also to the Intelligent Agent concept. More specifically:
— The DF in FIPA framework corresponds to the `MAFAgentFinder` functions in MASIF
— The AMS in FIPA corresponds to the `MAFAgent` System functions in MASIF
— The other MASIF agent system component/services (offered to the MA via places) become specific Intelligent Agent services within a FIPA Agent Platform (e.g. Wrapper, Resource Broker, and other intelligent agents for specific applications).
• The concept of *region* in OMG MASIF is used to logically group the physical agent systems and their services (MASIF service and non-MASIF services) for a single local authority, while the *domain* concept in FIPA aims at logically grouping the Intelligent Agents services. We can conclude that region has some similarity to the domain concept in FIPA IA framework.

It is obvious that both IA and MA religions have their strengths and weaknesses in the versatile application fields for the agent technology. Therefore, it is very likely that both paradigms will converge in the near future. A more detailed discussion of these aspects can be found in Zhang (1998).

## 6.3    Mobile Agent Platforms

The boom of research activities related to Mobile Agent Platforms started in the mid nineties, motivated by several advantages promised by this new technology, e.g. asynchronous task execution, reduction of network traffic, robustness, distributed task processing, and flexible on-demand service provision (Chess, 1998). Many research labs and manufacturers were involved in the development of various agent platforms, built on top of different operating systems, based on different programming languages and technologies. Even new languages have been realized, exclusively designed for the support of MA (e.g. TeleScript). However, within the last few years, common trends can be noticed: interpreter-based programming languages, particularly Java, are forming the basis for most of today's agent platforms. Additionally, several approaches are associated with the integration of Mobile Agents and RPC-based middleware like CORBA. As we have seen in the previous section, agent standards are paving the way towards such integrated platforms as outlined in Zhang (1998). In the following we will look closer to the capabilities provided by state of the art MAP as described in Breugst (1998a).

### 6.3.1   Common mobile agent platform concepts and capabilities

In the course of time, some Java based MAPs gained a certain level of acceptance, and represent the state of the art in MAT, namely *Aglets Workbench* (AWB) from IBM Japan, Grasshopper from IKV++, Odyssey from General Magic, and Voyager from ObjectSpace. In contrast to a more detailed description and comparison of these platforms, which can be found in Perdikeas (1999), we will provide a brief overview of the more general/common capabilities of these platforms.

In general, each Agent Platform comes with a class library that allows a programmer to develop his own agents and applications, subclassing when appropriate an already existing class. All of them make extensive use of Java's *Object Oriented* (OO) characteristics to assist the development process and exploit fully the intrinsic mobility properties of Java. The most important abstractions one needs to familiarize with, in order to gain an understanding of their internal mechanics are given above.

### 6.3.1.1    Agents

In each platform a base class providing the fundamental 'agent' behavior, exists. The developers can then subclass this class to add problem specific functionality. In some platforms this base class is used for all agents (static or mobile), while in others there are two separate classes. Programmatically the platform's runtime components that constitute the hosts to which agents migrate 'see' instances of this base class. The base class therefore provides methods that are called directly by the runtime components. In all the platforms, the host and the agents that are located in it are residing in the same address space, i.e. in the same *Java Virtual Machine* (JVM).

When the developer extends the base class to implement his agents, more methods are naturally added, but it is usually the case that the hosts do not perceive these new methods (i.e. the host-agent interaction is basically carried out by means of the base class' declared methods). Of all the base class methods there is one of particular importance that is often called, quite appropriately, the 'live' method. When an agent is for the first time instantiated in a host (i.e. in the virtual machine to which the host components reside), the host components create a separate thread of execution and invoke the `live` method on the newly instantiated agent. From that point on the agent is free to create new agents or spawn new threads if it wishes. The only restrictions that it has to observe (relative to what is the case for the host components) are those imposed by the Java security manager. When the agent finishes its main thread (exits from the `live` method) and all the other threads that were spawned from it, it remains as an object in the memory. The Java garbage collection cannot at that point step in simply because the host's runtime components continue to hold references to the agent. In general, this problem is difficult to solve. Even if the garbage collector were able to discard the agent it may not be the right thing to do. For instance, remotely located agents may wish to communicate with that agent at a certain point in the future. This naturally relates to the communication mechanisms used by a platform and the means that are available to determine whether such 'remote references' exist.

For the time being, it is sufficient to say that certain platforms (e.g. Voyager) provide distributed garbage collection mechanisms that deal with that problem. Others (e.g. Aglets) adopt a more conservative approach by temporarily (and transparently to the developer) deactivating the agents that are idle for a certain amount of time. When a communication request arrives for the deactivated agent, the agent is read from persistence storage back into memory.

### 6.3.1.2    Hosts

This is the term[10] used by the platforms' documentation to collectively refer to the components of the framework that must be installed at a computer node, and that provide the necessary runtime environment for the agents to execute. The framework is responsible for invoking certain methods on the agents and for providing required services to them (e.g. migration, serialization and communication). Generally, the platforms offer to the

---

[10] Also appearing as environments or agencies or contexts or servers.

developer rather limited ability to extend the framework behavior (for instance, by granting him access to certain framework components or by defining some extension mechanisms). This is almost never constraining, since the development process is mostly focused on implementing the desired agents, and not on changing the behavior of the platforms, which are, in any case, generic components that can pose very little limitations by themselves.

Regarding the agencies, however, what is in most cases lacking is a clear – available to the developer – interface to the communication protocols used by a platform. One can easily envision situations where a developer wishes to use a platform, albeit with a different transport protocol. In these cases it would be useful if the developer had the ability to 'plug' a different communication component into the platform. Since it is between hosts that agents migrate, the hosts addresses appear as arguments in the invocation of the migration operations. This brings up the issue of which addressing scheme is used to identify hosts. In most platforms there is no host addressing scheme as such, but simply hosts are identified by the IP address of the node to which they reside. It would be useful if a different naming scheme was used to identify the hosts – to reflect the fact that hosts are at a higher level of abstraction than the actual computer nodes on which they reside.

Similar to the interface provided by the agents to the hosts (to help the latter manage the former), hosts also export an interface to their agents. Agents use that interface for accessing the communication facilities offered by the hosts, or to initiate transport operations. In some platforms, developer code does actually get references to the host underneath, while in others developers invoke methods on the base class (whose implementation naturally makes the actual call to the framework). Clearly, there is no real distinction between these two cases. What is important, however, is whether this interface is offered natively (i.e. as a Java interface) or as a CORBA interface to be accessed remotely. The MASIF specification mandates that at least two (rather terse) interfaces should be available for remote access.

### 6.3.1.3    Places

Places serve for the grouping of similar functionality. For instance a 'post place' may comprise services for sending e-mails, etc. At the present versions of the platforms, however, there is no extra functionality associated with a place (i.e. an agent does not enjoy any enhanced messaging capabilities by virtue of residing in the post place). It therefore appears that places serve only a logical 'categorization' purpose. Since the concept appears however at the MASIF standard, platforms that want to be MASIF-compliant have to support it. Where they appear, places serve as the ultimate destination points for the migration operations.

### 6.3.1.4    Entry points

This is not a concept that is related to the MA paradigm itself. Rather it is a necessary mechanism to circumvent a problem related to a characteristic of the JVM. Specifically, the JVM is unable to save the execution stack of a thread. So, in the event of an agent migration only the class bytecodes, and the instance's data can be transported to the remote host – but not the execution stack. The implication of this is that when, an agent invokes a move method (or what other name the migration method might have), execution at the destination will not continue from the next code statement, but instead a public method will be called that will allow the agent to continue processing on its own thread.

This method is appropriately called the 'entry point'. So agents generally have to explicitly save necessary state information to member variables allowing the entry point method to examine these variables and proceed depending on the state that the computation is in. Platforms may have one or multiple entry points. In the latter case, the developer has the option of specifying the method that should be invoked upon re-instantiation of the agent at the remote host.

### 6.3.1.5   Agent migration

Having described *agents*, *hosts* and *entry points*, it is now possible to see at a conceptually high level the migration of an agent performed between two hosts. Figure 6.6 provides a schematic illustration for that.

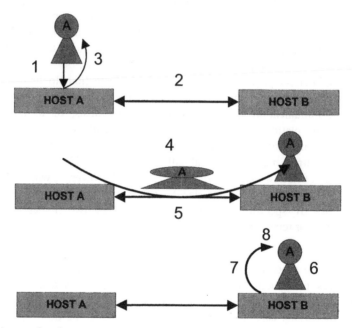

**Figure 6.6**   Agent migration

The steps associated with the migration procedure are:

1. The agent instructs host A to transfer it to host B. This can be done either by obtaining a reference to an object representing the hosts' capabilities and then by invoking the 'migration' method on the host, or by invoking a superclass 'migration' method (whose implementation will ultimately do the same). Either of the two approaches can be taken by the platform. At that point the call to the migration method blocks, and in the event of a successful migration it will be the last statement to be executed in host's A virtual machine.

2. Host A contacts host B and they make a dialogue over the migration to be performed. Issues to be resolved at that point include basically security considerations (whether the agent is allowed to migrate or whether host B can accept it), authentication/encryption (negotiation of permissions that the agent will have upon arriving at host B, exchange of

public keys or agreement of the secret key to be used for encoding the agent to be transported – in the case of symmetric encryption algorithms).

3. When that preliminary dialogue is over and host A has determined that the migration operation can proceed, it suspends execution of the agent (the execution had not gone any further than the move call) and serializes it into an array of bytes to be transported through the network.

4. The agents bytecode (i.e. its class definition) are transferred to host B.

5. The agent's serialized form (which comprises its member variables the state of which have been saved), is transferred to host B. Either one or both of the transfers can be encrypted using the keys that were negotiated during step 2 for further security.

6. When host B has the agent's bytecodes and data (the serialized form of its member variables) available locally it can, using the bytecodes, create an instance of the agent's class and using the saved state of the member variables, initialize it to the state it had just prior to the move call.

7. Host B creates a new thread.

8. In that thread, host B invokes the agent's entry point method.

### 6.3.1.6   Itineraries

Itineraries appear in most platforms. In most cases they are utility classes that help to better organize an agent's travel plan. Their use is seldom mandatory. From a programmatic point of view, it is not necessary that information about an agent's itinerary is kept as a separated data structure – it can be mingled with the code as well. Even in platforms that do not provide such a utility class, the developer can easily implement one.

### 6.3.1.7   Agent communication – proxies

A proxy is essentially a representative that a MA leaves when it migrates from a node and can be used to forward messages or method invocations to that MA in a location independent manner. Not all platforms implement proxies in the same way. The most crucial difference is whether or not arbitrary methods of an agent can be called remotely through a proxy. Platforms that support this functionality provide a utility that parses a Mobile Agent's class and creates the corresponding proxy, similar to the *Remote Method Invocation* (RMI) paradigm or to the CORBA's automatically generated stubs and skeletons. In platforms where arbitrary RMI through a proxy is not supported the proxy object provides only a uniform, generic method to send messages (e.g. sendMessage in Aglets) and therefore no proxy-generation utility is required. In this latter case, the generic method contains a generic argument so it is easy for the developer to implement a scheme whereby the agent examines some field of the generic argument and then calls a more specific method to carry on. In this way, the developer can mimic the arbitrary method feature.

The idea is that information concerning the actual location of an agent is not kept by the agents that are communicating but by their proxies. Other agents obtain references to the proxies and use these references to communicate in a location transparent manner with the actual agents. It is the responsibility of the proxies to be kept up to date with the actual location of the agent they represent and to route the messages they receive accordingly. In some platforms this location transparency is not so fully supported; for instance it may be the case that the developer is responsible for re-obtaining new proxies when an agent has migrated. Clearly this is a rather small inconvenience, but it nevertheless falls short of providing real location transparency. Other platforms support this feature more seamlessly by delegating that responsibility to the proxies that are then automatically and without developer's intervention refreshing the information they hold. In these cases, the developer has only to obtain a proxy for each agent he wishes to interact, and can use it for the entire lifetime of the 'represented' agent.

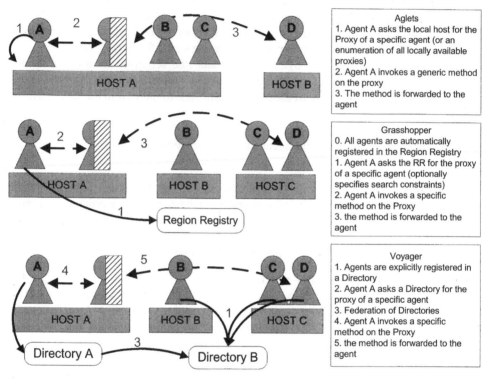

**Figure 6.7** Agent communication via proxies

Central to the discussion of proxies is the issue of how agents obtain references to them. Generally unique identifiers are associated with each agent, and interfaces that can return a corresponding proxy are provided, taking as input the identifier for an agent. So the key issue actually becomes how the platforms allow agents to obtain identifiers of other agents with which they wish to communicate. Some platforms offer very rudimentary interfaces, providing enumerations for only locally available agents and do not implement any mechanism to allow a developer to retrieve the identifier of a remote agent. It is the developers' responsibility to implement appropriate mechanisms. Other platforms offer a Directory Service and allow a program to query the directory for specific agents using certain criteria. Agents can be implicitly registered with that directory service, or they may need to be explicitly registered (the CORBA Naming Service model). Depending on the kind of the specified criteria that Directory Service can approximate the functionality of the CORBA Naming Service or that of a Trading Service. For instance, the Voyager platform expects the agents to explicitly register themselves under a string. This corresponds to the Naming Service paradigm. In addition to this functionality, the Grasshopper platform allows queries against methods offered by the agents which is more akin to the Trading Service. In general, all these queries return an agent identifier (or a set of agent identifiers corresponding to all agents that fitted the query for that matter) which can then be used to obtain a proxy. Another point of comparison is whether the directory service is centralized to a single location or a federation of directory services can be supported. Figure 6.7 illustrates how proxies can be obtained under different directory models.

Having acquired a proxy to a remote agent, an agent (or maybe a non-agentized piece of software) can use that proxy to invoke methods on and generally engage in communication with it. In that respect proxies function like remote references in CORBA. All platforms that implement proxies offer the usual blocking semantics. Alternatively, some platforms offer asynchronous messages, which can fall into several categories. One of them is future reply messages where the method returns immediately a `FutureReply` object which can be polled later for results. Instead of continuously polling the `FutureReply` object an agent (or any other piece of software) can invoke a blocking method on it that will return the actual result when this is available. The difference with the 'normal' blocking method call is that invoking the method and retrieving the result can happen at two different places or two different threads. Alternatively a subscribing mode is available where the agent can subscribe to the `FutureReply` object and have a callback method invoked on it when a result is available. Mobile Agents Platforms usually implement their communication facilities on top of other application level infrastructures such as CORBA or Remote Method Invocation and in certain occasions directly on top of plain TCP/IP.

### 6.3.1.8   Persistence

MAPs offer a way for agents to suspend their execution, be serialized into a persistence medium and restored back after a period of time to continue their operations. After resuming an agent, an entry point method has to be called as in the case of migration. Platforms support persistence in different ways. Some allow an agent to suspend itself only for a certain (usually limited) time interval. Others can allow suspending an agent for an arbitrary time period and/or support wakening it due to asynchronous events that may occur. In all these cases persistence is under developer's control. A third option is to implement mechanisms that automatically and transparently to the developer suspend an idle agent in order to save resources. This particular use of persistence is more of a platform improvement than a class library feature at the developer's disposal.

### 6.3.1.9   Security

In general, security concerns in MAT revolve around the following: protect hosts from malicious agents, protect agents from malicious hosts, enjoy secure agent-to-agent communications, ensure that an agent is not tampered with while migrating and that its contents cannot be determined by potential eavesdroppers. Solutions to the above issues use public or private key encryption techniques depending on the criticality of the application and the level of complexity that it warrants.

Approaches to protecting hosts from malicious agents can for instance mandate that a host does not receive an agent unless the agent's identity or the agent's authors identity can be determined and verified that the agent or its author are trusted. Public key schemes and trusted third party certificates can be used for that purpose. A similar approach is to trust an agent, if the administrative region to which it had previously been roaming is considered secure. For not trusted agents, migration to a host can be altogether denied or strict restrictions on its operations can be imposed by a security manager. Protecting agents from malicious hosts could be similarly guaranteed, although we are not aware of any platform allowing agents to assess the 'level of trust' they can afford to assign to a host. As MAT moves to actual commercial deployments, we expect some mechanisms of that sort to appear. One has to recognize, however, that the problem of protecting agents against malicious hosts is more difficult to tackle since a hosts needs complete access to an agents bytecodes in order to execute it.

Secure communications can likewise be established by public or secret key schemas. The problem of verifying that an agent was not tackled while *enroute* is also amenable to solutions that either encrypt an agent, guaranteeing that its bytecodes can't be retrieved in any usable form by eavesdroppers or that couple agents with *Message Authentication Codes* (MACs) – hash functions whose value will most likely change if the agent is modified. Certain MAPs have taken the approach of not providing any application-level security enhancements as such, but by resting on secure transport-level protocols (e.g. *Secure Socket Layer* – SSL).

## 6.3.2    Integrating MAT platforms and DOT platforms for applications

CORBA has been established as an important standard, enhancing the original client-server architectures by allowing relatively free and transparent distribution of service functionality. Initially, MAT has been regarded as a radically different way of implementation of client/server computing. But in specific situations remote client-server communications is more appropriate than shipping Mobile Agents back and forth. Consequently, there is today a common understanding that MAT should be seen as an enhancement of DOT rather than a replacement. Due to its benefits, such as dynamic, on-demand provision and distribution of services, reduction of network traffic and the reduction of dependence regarding server failures, various problems and inefficiencies of today's client/server architectures can be handled by means of this new paradigm. Thus both technologies offer advantages in different aspects, the comparison of which provides the maximum flexibility in application design.

The obvious need for integrated solutions have resulted in the development of CORBA-based MATs as described in Krause (1997) and Breugst (1998). One of these is the Grasshopper Agent Platform (Grasshopper, 1999;Bäumer, 1999). These platforms are combining the benefits of both client-server and MAT, and are eliminating or at least minimizing the problems that rise if one of these techniques, i.e. DOT and MAT, is used as 'stand-alone' solution.

In the following we briefly describe the general blueprint for such an integrated platform, and look at the Grasshopper Agent Platform as one particular example in some more detail.

### 6.3.2.1    General architecture principles
Figure 6.8 shows the blueprint of such a CORBA-based MAP, which enables the establishment of a *Distributed Agent Environment* (DAE) on top of a CORBA-based *Distributed Processing Environment* (DPE). The basic idea of such a platform is to use CORBA for both the modeling of and communication between components inside an agency, i.e. the actual runtime environment for MA. Thus agents, providing CORBA interfaces (i.e. IDL interfaces), communicate with the core services of the agencies via

CORBA. Furthermore, agent transport as well as further interactions between distinguished DAE and non-DAE components are also performed via CORBA. In this way, existing CORBA (legacy) services, like a CORBA trading, naming, or event service can be used to realize and enhance the platform functionality in a very comfortable manner. By providing the CORBA interfaces specified by the OMG MASIF standard, interoperability to other agent platforms can be achieved. In this way, an integration of both Mobile Agents and remote/cross network client-server communication is achieved. This allows to evaluate for each application both possibilities, i.e. to decide whether to use agent migration plus local interactions or to use remote interactions instead.

To support various application areas with specific needs, the functionality of such a target CORBA-based MAP is separated into the core functionality of the platform and additional capabilities. The core agency comprises only those capabilities that are inevitably necessary for the maintenance of the agency itself, as well as for the management, execution and migration of MA. Additional, application-dependent functionality is realized by means of modular building blocks that can be added to a core agency, and thereby enhance the platform capabilities, depending on concrete application scenarios. Examples of building blocks are adapter services for the access of telecommunication resources or sophisticated communication services (e.g. *Intelligent Network Application Protocol* (INAP) or *Common Management Information Protocol* (CMIP) protocol stacks). Since this platform is based on CORBA, already existing INAP-CORBA and CMIP-CORBA gateways can be (re)used for that, which illustrates furthermore the benefits of the integrated approach.

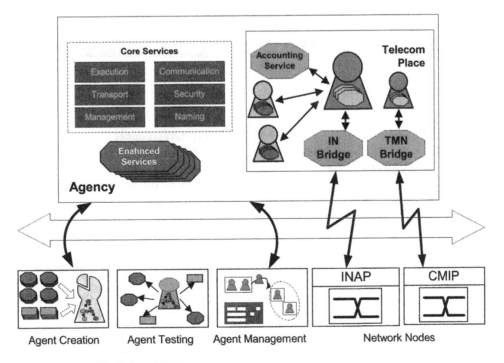

**Figure 6.8**  A CORBA-based MAP

Furthermore, the following tools are additionally required:

- An agent creation environment, enabling the 'plug and play' composition of mobile agents out of reusable functional building blocks.
- An agent testing environment, allowing for the simulation of the whole distributed environment by means of a single agency. The entire execution of an agent can be simulated 'locally' without endangering any real resources.
- Finally, a (graphical) agent management tool, enabling the monitoring and control of agents and agencies in the scope of one or more regions.

Through the concerted deployment of such a CORBA-based MAP in all relevant network nodes constituting a telecommunications network, a homogenous agent environment is established on top of an underlying DPE. In this environment agents can roam and access local and remote agents and services in a transparent manner (see Figure 6.9).

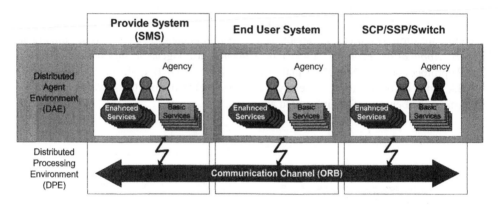

**Figure 6.9**   Roaming agents can access local and remote services

### 6.3.2.2    An example CORBA-based MAT platform – Grasshopper

Grasshopper, which has been developed jointly by TUB OKS, GMD FOKUS and IKV++ GmbH, is a Mobile Agent development and runtime platform that is built on top of a Distributed Processing Environment. Thus, an integration of the traditional client-server paradigm and Mobile Agent Technology is achieved. Grasshopper is implemented in Java, based on the Java 2 specification. Most importantly, Grasshopper has been designed to be conformant with the first industry Mobile Agent standard, namely the Object Management Group's MASIF.

In principle, Grasshopper realizes a DAE. That DAE is composed of regions, places, agencies and different types of agents. A place provides a logical grouping of functionality inside an agency. The region concept facilitates the management of the distributed components (agencies, places, and agents) in the Grasshopper environment. Agencies as well as their places can be associated with a specific region by registering them within the accompanying region registry. All agents that are currently hosted by those agencies will also be automatically registered by the region registry. If an agent moves to another location, the corresponding registry information is automatically updated. Figure 6.10 depicts an abstract view of these entities.

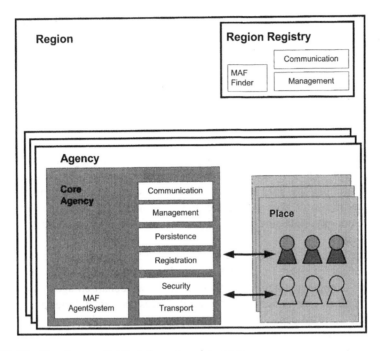

**Figure 6.10**   The Grasshopper distributed agent environment

Two types of agents are distinguished in Grasshopper: Mobile Agents and *Stationary Agents* (SA). The actual runtime environment for both Mobile and Stationary Agents is an agency: on each host at least one agency has to run to support the execution of agents. A Grasshopper agency consists of two parts: the core agency and one or more places. Core Agencies represent the minimal functionality required by an agency in order to support the execution of agents. The following services are provided by a Grasshopper core agency:

- Communication Service. This service is responsible for all remote interactions that take place between the distributed components of Grasshopper, such as location-transparent inter-agent communication, agent transport, and the localization of agents by means of the region registry. All interactions can be performed via the Internet Inter-ORB Protocol, Java's RMI, or plain socket connections. Optionally, RMI and plain socket connections can be protected by means of the SSL, which is the *defacto* standard Internet security protocol. The communication service supports synchronous and asynchronous communication, multicast communication, as well as dynamic method invocation. As an alternative to the communication service, Grasshopper can use its OMG MASIF-compliant CORBA interfaces for remote interactions. For this purpose, each agency provides the interface MAFAgentSystem, and the region registries provide the interface MAFFinder.

- Registration Service. Each agency must be able to know about all agents and places currently hosted, on the one hand for external management purposes and, on the other hand, in order to deliver information about registered entities to hosted agents. Furthermore, the registration service of each agency is connected to the region registry which maintains information of agents, agencies and places in the scope of an entire region.
- Management Service. The management services allow the monitoring and control of agents and places of an agency by (human) users. It is possible, among others, to create, remove, suspend and resume agents, services, and places, in order to get information about specific agents and services, to list all agents residing in a specific place, and to list all places of an agency.
- Security Service. Grasshopper supports two security mechanisms: external and internal security.
  — External security protects remote interactions between the distributed Grasshopper components, i.e. between agencies and region registries. For this purpose, X.509 certificates and the SSL are used. SSL is an industry standard protocol that makes substantial use of both symmetric and asymmetric cryptography. By using SSL, confidentiality, data integrity, and mutual authentication of both partners can be achieved.
  — Internal security protects agency resources from unauthorized access by agents. Besides, it is used to protect agents from each other. This is achieved by authenticating and authorizing the user on whose behalf an agent is executed. Due to the authentication/authorization results, access control policies are activated. The internal security capabilities of Grasshopper are mainly based on Java security mechanisms.
- Persistence Service. The Grasshopper persistence service enables the storage of agents and places (the internal information maintained inside these components) on a persistent medium. In this way, it is possible to recover agents or places when needed, e.g. when an agency is restarted after a system crash.

## 6.4    Mobile Agents in Telecommunications

Today the agent paradigm and related agent technologies are deployed in nearly all application domains. Probably the most interesting area is the telecommunications environment, where agents are considered key for the implementation of flexible and scalable solutions for an open telecommunication services market within the emerging information society (Magedanz, 1996b). The heterogeneity, distribution, scale and dynamic nature of the emerging open telecommunication environments, sometimes referred to as 'information infrastructure', are calling for new paradigms for the control and management of the open resources. In this context, the agent-based technology seems to offer a promising solution to cope with the complexity of this environment. Within such an open environment, an agent-based solution can:

- Reduce the requirement of traffic load and the availability on the underlying networks (via the autonomy and asynchronous operations of the agents).

- Reduce the requirement on customer intelligence during the installation, operation and maintenance of the resources (via the intelligence and autonomy of the agents).
- Enable 'on demand' provision of customized services (via dynamic agent downloading from the provider system to the customer system and further on back to the provider system or directly to the resources).
- Increase the flexibility, reusability and effectiveness of the software-based problem solutions.
- Allow for a more decentralized realization of service control and management software, by means of bringing the control or management agent as close as possible or even onto the resources.

Therefore, a lot of international research is undertaken within ESPRIT, ACTS and the new IST frameworks of the European Commission (see (ESPRIT, 1998) and (ACTS, 1998)).

Buzzwords such as 'Management by Delegation' (MbD) in the context of distributed network management, 'Intelligence on demand' in the context of distributed intelligent networks, and most recently 'Active Networking', which considers the dynamic programmability of the (internet) network nodes for both control and management applications, indicate the general direction of network evolution. Furthermore, MAT is also considered in the context of mobile communications, where particularly Mobile Agents are used to enable migrating user agents following the mobile users across network boundaries and dynamic adaptation of services in the context of varying access network capabilities and the *Virtual Home Environment* (VHE) realization (Breugst, 1998c). In the following we look in some more detail at the first three application domains.

For more comprehensive overviews, readers are referred elsewhere (Albayrak, 1998; Hayzelden, 1999;Karmouch, 1998;Magedanz, 1999, 2000).

### 6.4.1    Management by delegation – MA-based network management services

Current network management systems are typically designed according to a centralized network management paradigm characterized by a low degree of flexibility and re-configurability. Management interactions are based on a centralized, client-server model, where a central station (manager) collects, aggregates and processes data retrieved from physically distributed servers (management agents). Widely deployed network management standards, such as the *Simple Network Management Protocol* (SNMP) of the TCP/IP protocol suite or the CMIP of the OSI world, are designed according to this rigid centralized model. Within these protocols, physical resources are represented by managed objects. Collections of managed objects are grouped into tree-structured *Management Information Bases* (MIBs).

The centralized approach in network management presents severe efficiency and scalability limitations: the process of data collection and analysis typically involves massive transfers of management data causing considerable strain on network throughput and processing bottlenecks at the manager host. All these problems suggest distribution of management intelligence as a rational approach to overcome the limitations of centralized network management.

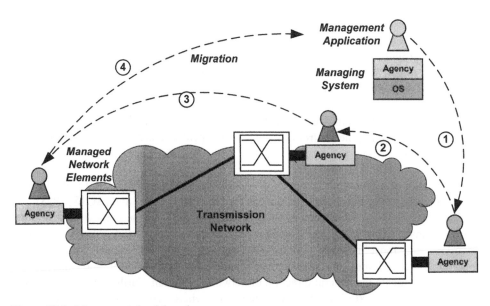

**Figure 6.11**   Management by delegation

In the light of agent technology, the concept of 'Management by Delegation' (Yemini, 1991;Goldszmidt, 1995, 1998) represents the first clear effort towards decentralization of management logic. This idea was expressed before developing the first systems based on mobile code. The initial approach was to download management scripts that were compiled and executed at the agent side. This means that the management station sends some commands to the remote device (instead of limiting itself to the request for MIB variables), delegating to the remote agent the actual execution. For that, remote devices had to include a so-called elastic process runtime support, which provided for new services and dynamically extension of the ones present in the device.

Finally, the advent of Java and related MA platforms has made this task significantly simpler. The idea of management distribution is taken further by solutions that exploit MA, which provide a powerful software interaction paradigm that allows code migration between hosts for remote execution. The MA ability to store the state during the movements allows developing applications in which the agent moves from a device to another, performing the management functions required locally. Besides, the agent can be assigned the task to decide where and when to migrate according to its current state, thus reducing the interaction with the network management system, and making processing more distributed. This ability has attracted much attention to MAT, with several MAPs proposed for network management applications (Eckardt, 1996a;Zapf, 1999;Bieszczad, 1999).

### 6.4.2   *Intelligence on demand – MA-based intelligent network services*

The concept of Intelligence on Demand has been introduced in the midnineties as a projection of the Management by Delegation approach to Intelligent Networks. The concept promotes the implementation of IN services through Mobile Agents, and thereby the dynamic deployment and distribution of these services directly to the Service Control Points, the Service Switching Points and even the end user systems (Magedanz, 1996a, 1997, 1998). Figure 6.12 illustrates the main idea. The main motivation is to overcome the intrinsic bottlenecks of the centralized IN architecture by distributing the services to the switches. Therefore, this concept can be seen as the starting point for the later considerations of Chapter 7. We therefore skip any further considerations at this place.

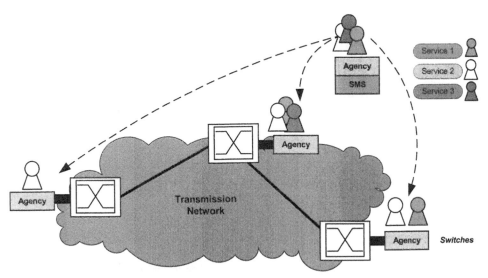

**Figure 6.12**   Intelligence on demand concept

### 6.4.3   Active networking – MA-based services on the internet

The notion of Active Networks describes a new discipline in Internet research, where the network moves from the old 'store and forward' paradigm to a new 'store, compute, and forward'. The basic idea of *'Active Networks'* (Tennenhouse, 1997; IEEE, 1998) is the movement of service code, which has been traditionally placed outside the transport network, directly to the network's switching nodes, i.e. in the so-called active routers. This movement of service code should be possible in a highly dynamic manner. This allows the automated, flexible, and customized provision of services in a highly distributed way, thereby enabling better service performance and optimized control and management of transport capabilities.

Two enabling technologies are key for this: *programmable switches*, which provide flexibility in the design of connectivity control applications, and *Mobile Code Systems/Mobile Agent Platforms*, which enable the dynamic downloading and movement of service code to specific network nodes. In contrast to the *integrated approach* of active network research, where the transmitted packets (so-called capsules) contain besides the data, program fragments responsible for processing the data at the switches, the *discrete approach* of active networking, foresees separate service deployment (i.e. outband) prior to service processing. This is very much in line with the previous considerations of Management by Delegation and Intelligence on Demand (Breugst, 1998b). Looking at current Active Network applications, one can identify primarily active multicasting, network management, and media format conversion.

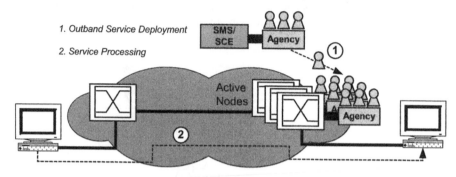

**Figure 6.13**   Active networks – active node approach

Figure 6.13 depicts the general idea of dynamic programming of active switches/routers by means of MA. Note that these are deployed via a dedicated network (outband) prior to the actual service execution.

## References

ACTS (1998) CLIMATE Home Page: http://www.fokus.gmd.de/cc/ima/climate/.

Albayrak, S. (1998) *Intelligent Agents for Telecommunications Applications*. ISNN: 90-5199-295-5, IOS Press, 1998.

Bäumer, C. and Magedanz, T. (1999) Grasshopper – An Agent Platform for Implementing Active Telecommunication Networks. *3rd International Workshop on Intelligent Agents for Telecommunication Applications (IATA)*, Stockholm, Sweden, August 1999.

Bieszczad, A., Pagurek, B. and White, T. (1999) Mobile Agents for Network Management. *IEEE Communications Surveys*, 1(1), http://www.comsoc.org/pubs/ surveys.

Borenstein, N.S. (1992) Computational Mail as Network Infrastructure for Computer-Supported Cooperative Work, *CSCW'92 Proceedings*, Toronto.

Borenstein, N.S. (1994) EMail with a Mind of its own: The Safe-Tcl Language for Enabled Mail. ULPAA, Barcelona.

Bradshaw, J.M. (1997) *Software Agents*. MIT Press, ISBN 0-262-52234-9.

Breugst, M. and Magedanz, T. (1998b) Mobile Agents – Enabling Technology for Active Intelligent Networks. *IEEE Network Magazine*, 12(3), 53–60.

Breugst, M. and Magedanz, T. (1998c) GRASSHOPPER – A Mobile Agent Platform for IN Environments. ISBN: 0-7803-4905-9, IEEE Catalog No.: 98TH8364, *IEEE IN Workshop 1998*, Bordeaux, France, May 10–12 1998, 279–290.

Breugst, M., Hagen, L. and Magedanz, T. (1998) Impacts of Mobile Agent Technology on Mobile Communication System Evolution. *IEEE Personal Communications Magazine*, 5(4), 56–69.

Breugst, M. and Magedanz, T. (1998a) On the Usage of Standard Mobile Agent Platforms in Telecommunication Environments. *LNCS 1430, Intelligence in Services and Networks:*

*Technologies for Ubiquitous Telecom Services*, S. Trigila et al. (Eds.), 275–286, ISBN: 3-540-64598-5, Springer-Verlag.

Chess, D. et al. (1995) Itinerant Agents for Mobile Computing. *IEEE Personal Communications Magazine*, **2**(5), 34–59.

Choy, S., Breugst, M. and Magedanz, T. (1999) Beyond Mobile Agents with CORBA – Towards Mobile CORBA Objects. *6th ACTS Conference on Intelligence in Services and Networks (IS&N)*, pp. 168-180, H. Zuidweg et.al. (Eds.), IS&N 99, LNCS 1597, ISBN: 3-540-65895-5, Springer-Verlag, 1999

Chess, D., Harrison, C.G. and Kershenbaum, A. (1998) Mobile agents: Are they a good idea?. *LNCS 1419*, G. Vigna (Eds.), *Mobile Agents and Security*, 25–47, Springer-Verlag, 1998.

*Communications of the ACM* (1994) Intelligent Agents, **37**(7), July 1994.

Eckardt, T. and Magedanz, T. (1996a) Mobile Software Agents: A new Paradigm for Telecommunications Management, in *Proceedings of IEEE/IFIP Network Operations and Management Symposium (NOMS)*, 360-369, IEEE Catalog No. 96CH35757, ISBN: 0-7803-2518-4, IEEE Press, 1996.

ESPRIT (1998), AgentLink Home Page: http://www.AgentLink.org/.

Etzioni, E. and Weld, D. (1994) A Softbot-Based Interface to the Internet. *Communications of the ACM*, **37**(7), 72–76, July.

Finin, T. et.al. (1994) KQML as an Agent Communication Language. *3rd International Conference on Information and Knowledge Management (CIKM'94)*, ACM Press.

FIPA (1998) – Part 2: Agent Communication Language (V.2.0).

Foundation for Intelligent Physical Agents (FIPA) Home page: http://www.fipa.org/

Fuggetta, A., Picco, G. P. and Vigna, G. (1998) Understanding Code Mobility. *IEEE Transactions on Software Engineering*, **24**(5), 342–362, May.

Ginsberg, M. (1991) Knowledge Interchange Format: the kif of death. *AI Magazine*, **5**(63).

Goldszmidt, G. and Yemini, Y. (1995) Distributed Management by Delegation. *15th International Conference on Distributed Computing Systems*, IEEE Computer Society, June 1995.

Goldszmidt, G. and Yemini, Y. (1998) Delegated Agents for Network Management. *IEEE Communications Magazine*, **36**(3), March.

Grasshopper (1999) IKV++ GmbH, Grasshopper Home page: http://www.ikv.de/products/-grasshopper

Guilfoyle, C. and Warner, E. (1994) *Intelligent Agents: the New Revolution in Software*. Technical Report, OVUM Limited.

Harrison, C.G. et al. (1995) Mobile Agents: Are they a good Idea?. *IBM Research Report*, RC 19887.

Hayzelden, A.L.G. and Bigham, J. (Eds.), (1999) *Software Agents for Future Communication Systems*. ISBN: 3-540-65578-6, Springer Verlag.

*IEEE Network Magazine* (1998), **12**(3), Special Issue on Active and Programmable Networks.

ISO/IEC 9596, Information Technology, Open Systems Interconnection, Common Management Information Protocol (CMIP) – Part 1: Specification, Geneva, Switzerland, 1991.

Jennings, N.R. and Wooldridge, M.J. (Ed). (1998) *Agent Technology: Foundations, Applications and Markets*. Springer-Verlag 1998, ISBN 3-540-63591-2.

Karmouch, A. (Ed.) (1998) Special Issue on Mobile Software Agents for Telecommunications. *IEEE Communications Magazine*, **36**(7).

Krause, S. and Magedanz, T. (1997) MAGNA – A DPE-Based Platform for Mobile Agents in Electronic Service Markets. *IEEE International Symposium on Autonomous Decentralized Systems (ISADS)*, 93–102, IEEE Catalog No.97TB100111, ISBN: 0-8186-7785-6, IEEE Press.

Maes, P (1994) Agents that reduce work and information overload. *Communications of ACM*, **37**(7), 31–40.

Magedanz, T. (1997) Mobile Agents – Basics, Technologies, and Applications (including Java and CORBA Integration). Invited Tutorial at *IEEE IN Workshop*, Colorado Springs, USA, May.

Magedanz, T. (1998) Mobile Agent based Service Provison in Intelligent Networks. In *Frontiers in Artificial Intelligence and Applications*, **36**, Intelligent Agents in Telecommunications Applications, S. Albayrak (Ed.), 94–104, ISBN 90 5199 295 5, IOS Press.

Magedanz, T. (Ed.) (1999) Special Issue on Mobile Agents in Intelligent Networks and Mobile Communication Systems. *Computer Networks Journal*, **31**(10), July.

Magedanz, T. and Glitho, R. (Eds.) (1999), Special Issue on Mobile Agent-based Network and Service Management. *Journal of Network and Service Management (JNSM)*, **7**(3).

Magedanz, T. and Karmouch, A. (Eds.) (2000) Special Issue on Mobile Agents for Telecommunication Applications. *Computer Communications*, **23**(8).

Magedanz, T. and Krause, K. (1997) Mobile Agents – Basics, Technologies, Standards, and Applications. Invited Tutorial at *1st International Workshop on Mobile Agents*, Berlin, Germany, April 1997.

Magedanz, T. and Popescu-Zeletin, R. (1996a) Towards Intelligence on Demand – On the Impacts of Intelligent Agents on IN. *Proceedings of 4th International Conference on Intelligent Networks (ICIN)*, Bordeaux, France, December 2–5, 30–35.

Magedanz, T.K., Rothermel, R. and Krause, K. (1996b) Intelligent Agents: An Emerging Technology for Next Generation Telecommunications? *Proceedings of IEEE INFOCOM '96*, 464–472, IEEE Catalog No. 96CB35887, ISBN: 0-8186-7292-7, IEEE Press.

Magedanz, T., Wade, W. and Donelly, W. (Eds.) (1999) Special Issue on Broadband Telecommunications Management. *Interoperable Communication Networks (ICON) Journal*, **2**(1), March.

Nwana, H.S. (1996) Software agents: an overview. *The Knowledge Engineering Review*, **11**(3), 205–244.

OMG CORBA (1995) Common Object Request Broker Architecture and Specification, Revision 2, August 1995.

OMG MASIF (1997) Mobile Agent System Interoperability Facility (MASIF) specification, November 1997, ftp://ftp.omg.org/pub/docs/orbos/97-10-05.pdf.

OMG: RFP 3 (1995) Common Facilities Request for Proposal OMG TC Document 95-11-3, November 1995, http://www.omg.org.

Perdikeas, M.K., Chatzipapadopoulos, F.G., Venieris, I.S. and Marino, G. (1999) Mobile Agent Standards and Available Platforms. *Computer Networks Journal*, Special Issue on Mobile Agents in Intelligent Networks and Mobile Communication Systems, **31**(10).

Reinhardt, A. (1994) The Network with Smarts. *BYTE Magazine*, 51–64, October.

Searle, J. (1969) Speech Acts – *An Essay in the Philosophy of Language*. Cambridge University Press, U.K.

Stamos, J.W. and Grifford, D.K. (1990) Implementting Remote Evaluation. *IEEE Transactions on Software Engineering*, **16**(7), 710–722.

Tennenhouse, D., Smith, J., Sincoskie, W. D. and Wetherall, D. (1997) A Survey of Active Network Research. *IEEE Communications Magazine*, January.

UMBC: *Knowledge Sharing Effort*, http://www.cs.umbc.edu/kse or http://www.cs.umbc.edu/agents.

White, J.E. (1994) Telescript Technology: The Foundation for the Electronic Marketplace. General Magic White Paper.

White, J. E. (1997) Mobile Agents. In Software Agents, Bradshaw, J. M. (Ed.) MIT Press, ISBN 0-262-52234-9. Chapter 20. 437–472.

Wooldridge, M. and Jennings, N.R. (1995) Intelligent Agents: Theory and Practice. *The Knowledge Engineering Review*, **10**(2), 115–152.

Yemini Y., Goldszmidt G. and Yemini S. (1991) Network Management by Delegation. *Proceedings of the 2ⁿᵈ International Symposium on Integrated Network Management*, 95–107.

Zapf, M., Herrmann, K. and Geihs, K. (1999) Decentralised SNMP Management with Mobile Agents. In *Integrated Network Management VI*, Sloman, Mazumdar, Lupu (Eds.), 623–635, IEEE.

Zhang, T., Magedanz, T. and Covaci, S. (1998) Mobile Agents vs. Intelligent Agents – Interoperability and Integration Issues. *4th International Symposium on Interworking*, Ottawa, Canada.

# Part III

Case Study: Distributed Intelligent Broadband Network

# 7

# Evolution towards a Distributed Intelligent Broadband Network

Nearing the end of this century, the *Intelligent Network* (IN) has become the most important technological basis for the rapid and uniform provision of advanced telecommunication services above an increasing number of different transmission networks. In line with the main objectives of the IN, namely to provide a network independent and service independent architecture, the IN can be regarded as a middleware platform for telecommunications environment, which is primarily based on the incorporation of state of the art information technology into the telecommunications world.

One fundamental principle of the IN and probably the key to its success is its inherent evolution capability, expressed also by the notion of standardized IN *Capability Sets* (CSs). This phased approach of standardization and thus evolution allows us, on the one hand, to extend the IN service capabilities and architecture as required by the market demands, i.e. by extending the scope of IN service provision from *Public Switched Telephone Network* (PSTN) and *Integrated Digital Services Networks* (ISDN) into *Broadband ISDN* (B-ISDN), *Public Land Mobile Networks* (PLMNs), and most recently, the Internet and *Voice over IP* (VoIP) networks. On the other hand, IN evolution takes into account the technological progress in the information technology world. Whereas the first IN architectures are based on the exploitation of the concepts from the distributed computing world, such as the remote procedure call paradigm, and the availability of common channel signaling networks in the seventies, newer IN architectures are driven by the emerging *Distributed Object Technologies* (DOT) and *Mobile Agent Technologies* (MAT) in the nineties (Magedanz, 1996b;Herzog, 1998).

This chapter discusses the evolution of the Intelligent Network towards a distributed IN, analyzing how the network itself is currently structured and identifying principles leading the migration towards the distributed IN. The important intermediate step of *Intelligent Broadband Networks* (IBN), which defines new resource models suitable for the control of advanced services, is also outlined.

*Object Oriented Software Technologies in Telecommunications*, Edited by I. Venieris, F. Zizza and T. Magedanz
© 2000 John Wiley & Sons Ltd

## 7.1     Basic Intelligent Network Principles

The Intelligent Network can be seen as the technological answer to the increasing customer demands and the liberalization of the telecommunications services market in the 1980s. It is an architectural framework for the rapid and uniform provision of an open set of telecom services on top of different switched telecommunication networks, such as the Public Switched Telephone Network. It is based on the establishment of connections between endpoints (terminals, typically telephone sets), i.e. telephone calls are established based on the underlying switched connection. IN represents a first step towards programmable switched telecommunication networks. Connections are created using large switching equipment; some – termed local exchanges – are directly connected to access lines (directly connecting a terminal), while others – termed transit exchanges – are only connected to trunks. In general, all the switching nodes provide the basic call processing capabilities required to establish a simple *Plain Old Telephone Service* (POTS) call. Any additional service capabilities going beyond the basic call handling, such as call forwarding, private numbering plan, incoming call screening, etc., are based on additional service logic and data, not included within the basic call processing capabilities of a switch. Traditionally, the realization of such supplementary services was switch-based, i.e. multiple switches hosted the service logic and data which made service provision expensive and time consuming due to the heterogeneity of the switching nodes constituting a network.

The IN has changed the means of telecom service implementation. Instead of replicating the service logic and data in all switching nodes, the basic idea of the IN is to move telecom services (i.e. the *network intelligence*) to 'centralized', high performance and reliable computing nodes in the network, and remotely control the establishment of call connections performed by the switches. Therefore, the switches have to be able to recognize when the remote node must be triggered within a call. Thus, within the IN approach service switching and service control have been separated, requiring the definition of the corresponding 'interfaces' between the switches, referred to as *Service Switching Points* (SSPs), and the 'intelligent nodes', known as *Service Control Points* (SCPs). The fundamental prerequisite for this approach is the 'real-time' connection between the SSPs and the SCPs, which became possible through the introduction of 'common channel signaling' networks relying on the principle of 'out of band signaling' in the 1970s. Today the international *Signaling System No. 7* (SS7) network represents the basis for all IN implementations supporting the *IN Application Protocol* (INAP) between SSP and SCP.

IN service deployment and management is realized through a *Service Management System* (SMS)[1], which does not interact via the SS7 network but via a data communications network with the IN network elements. In addition, the IN allows the introduction of new service capabilities, such as advanced user interactions, through the introduction of an additional network element connected directly to an SSP, referred to as *Intelligent Peripheral* (IP). The IP is shared between multiple SSPs, since it provides specialized resources, such as speech synthesis units, which could not be accommodated in each SSP due to cost reasons.

---

[1] Another important key concept of the IN is the introduction of reusable *Service Independent Building Blocks* (SIBs), which are used for the IN service creation within a *Service Creation Environment* (SCE). The created service logic and data will be passed after testing to the SMS for service deployment.

Figure 7.1 shows a general IN architecture. In the context of IN standardization taking place in ITU-T and ETSI, the architectural definition is based on the *IN Conceptual Model* (INCM), which separates a functional and a physical IN architecture in corresponding INCM planes[2]. This is very important for our subsequent considerations. In this context, the previously introduced IN entities, i.e. SSP, SCP, SMS, etc., correspond to *Physical Elements* (PEs), which contain one or multiple appropriate *Functional Elements* (FEs). The major defined FEs are the *Service Switching Function* (SSF), *Service Control Function* (SCF), *Service Data Function* (SDF), *Specialized Resource Function* (SRF) and the *Service Management Function* (SMF). The basic idea of this separation is to allow flexibility in IN implementation, i.e. an appropriate grouping of FEs into PEs in accordance with specific service and network requirements. In a traditional IN architecture an SSP contains an SSF, an SCP contains the SCF and SDF, and an IP contains the SRF. However, recent IN architectures are based on *Service Nodes* (SNs), which group SSF, SRF, SCF, and SDF into one physical entity, in order to provide IN services. Another element worth noting is a *Service Switching & Control Point* (SSCP) which groups SSF, SCF, and SDF.[3]

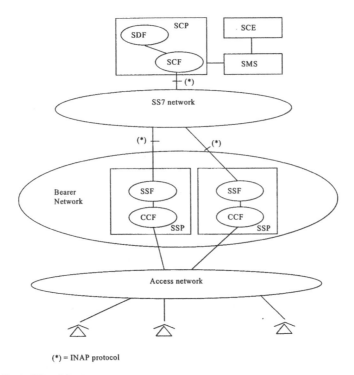

(*) = INAP protocol

**Figure 7.1** Basic IN architecture

---

[2] These INCM planes are called respectively the Distributed Functional Plane and the Physical Plane.

[3] Note that Figure 7.1 depicts an ideal IN architecture with one SCP controlling various SSPs. In reality, the SSPs do not necessarily have to be provided at the local exchange level, but can be implemented on a higher switching level. Nevertheless, there will be hundreds of SSPs, and for load sharing reasons multiple SCPs are providing different services in accordance with the number of IN service customers and network size. In addition, Service Nodes can provide dedicated services in specific geographic regions, e.g. big cities.

A closer look at the provision of IN services is now given. As the primary goal of IN services is to provide functionality beyond basic telephony services, IN services have to control network resources, i.e. the switching facilities. This approach requires the switches to be capable of providing visibility and control on detailed call events covering the complete process of a call. In plain telephony that means from picking up the receiver (off-hook) until the receiver is put down (on-hook), both for the originating and the terminating party. Hence, a corresponding model, known as the *Basic Call Model* (BCM), is required, which identifies all possible points in basic call processing, as seen in a switch, from which IN services can be invoked. To support multiple services by the IN architecture, both the basic call model and the protocol used between the SSP and the SCP should be ideally service independent. However, in reality this is impossible, and thus the pragmatic compromise adopted by the IN standardization was to define upward compatible stages of IN, known as Capability Sets. Each CS is defined by a specific set of IN benchmark services, supported by an appropriate call model and a corresponding IN protocol.

As the switch's basic call handling is modeled with high-level finite state machines, this model is also called *Basic Call State Model* (BCSM). The BCSM is built from two basic components: *Points in Call* (PICs) and *Detection Points* (DPs). PICs provide an external view of a call processing state or event to IN service logic. They are vendor independent, providing a standardized view of call processing behavior. DPs are placed between the PICs, and identify specific points in basic call processing at which specific events are detected and made visible to IN service logic, allowing for the transfer of control. The basic call processing may be suspended at a Detection Point, while waiting for instructions from a remote IN service logic. Whether IN service logic will be invoked or not depends on the consultation of specific trigger conditions. If the trigger conditions at a Detection Point are satisfied, then IN service logic processing will be initiated, otherwise (basic) call processing continues as if there had been no Detection Point (Magedanz, 1996a).

## 7.2    Intelligent Broadband Network

In the early nineties the broadband networking technology comes into play, giving a number of solutions to problems related to the simultaneous flow of information belonging to applications of different nature and scope. Broadband networks based on the *Asynchronous Transfer Mode* (ATM) principle can provide all those features required to integrate the transfer of different media over the same physical link with quality of service guarantees. Apart from the inherent property of ATM to provide switching and multiplexing of fixed size packets called cells by allocating information flows to different virtual connections, a number of associated mechanisms were developed to increase the efficiency of ATM-based broadband networks such as traffic shaping, policing and scheduling.

Although the traffic in ATM networks is organized in packets (cells), the ATM devices operate under the assumption that connections are either open or established through signaling. Looking at the latter case, which actually represents the way of assigning resources to user requests on demand, one could see that the currently used signaling systems cannot accommodate the complex associations required by new services at the communications level. In particular, both the *User to Network Interface* (UNI) and the *Network to Node Interface* (NNI) signaling systems (e.g. Q.2931, SS7) are more or less oriented to the support of simple configurations described as point to point, two party, single media associations. In fact, the rapid introduction of ATM both in the core public network and the local area pushed to the deployment of ready to use signaling systems. As such, the broadband network appeared as a powerful infrastructure in terms of physical communication resources, but unable to make proper use of them due to lack of mechanisms enabling the control of network resources. Similar to ISDN, one solution could be the introduction of IN technology, but now at the broadband era.

Three overlapping areas contribute to the concept of broadband IN, namely the services, the broadband network and the Intelligent Network area. The services area covers the design and implementation of new multimedia applications which, although in principle independent from the underlying network technology, in reality determine most of the decisions related to the network. The broadband network area covers the available infrastructure in terms of capabilities for traffic handling (user plane) and processing of call requests (control plane). The IN area covers the corresponding principles and methodology presented in Section 7.1. The common part of the three areas represents the developments necessary for the realization of the broadband IN, and will be discussed in the following using as a basis the description of the conventional IN.

The basic IN architecture depicted in Figure 7.1 is also valid for the case of broadband IN. The main differences are related to the enhanced role of the Functional Entities. The B-SSF (prefix B- is used to indicate broadband) should be able to present to the service logic a more complex association of resources as necessitated for the establishment of multiparty, multi-connection, multimedia calls. Also, the B-SSF should be able to handle bearer-unrelated control channels, or else deal with events irrelevant to any connection but which constitute instances of a user-service dialogue. We will further elaborate on this issue in the present chapter at the part discussing the *User Service Interaction* (USI) capability.

The B-SCF differs from the SCF mainly in the service logic and the protocol running between the B-SCF and the B-SSF, i.e. B-INAP. Both of them relate to the specific control features required for the underlying broadband network. For multimedia services the complexity in the design and implementation of IN service logic increases dramatically compared to the traditional IN. This is reflected not only in the number of *Service Independent Building Blocks* (SIBs)[4] required to model the service, but also in the service creation methodology which should guarantee a highly flexible system that can easily accommodate new services and network features. Traditional service creation platforms have proven to be unsuitable, while the object oriented design and implementation paradigm is currently the winning approach in modern platforms and service creation methodologies (Venieris, 1998).

The B-SRF differs significantly from its counterpart in the conventional IN. As this Functional Entity undertakes the communication between the non-IN aware user equipment and the Intelligent Network, it is natural that in the broadband context, the information exchanged will be massive and possibly of multimedia nature. Novel services like

---

[4] Each service logic program implementing an IN service is composed by a series of interconnected SIBs. Service execution always commences from the *Basic Call Processing* (BCP) SIB and proceeds to other SIBs through *Points of Initiation* (POIs). The service is terminated in a similar fashion when the execution returns to the BCP SIB through *Points of Return* (PORs). Following along this line, services can initially be represented by diagrams showing the interconnection between the various service stages and the *Basic Call State Model* (BCSM) through POIs and PORs.

broadband videoconference and interactive multimedia retrieval require enhanced functionality compared to the poor reproduction of voice messages. Video playback synchronized with high quality audio, hypertext documents, virtual reality and even voice recognition are some of the new media types that can be utilized in the broadband multimedia IN services. Therefore, the B-SRF should be able to host complex service logic programs and a number of advanced information transport capabilities, in order to communicate with a wide range of terminals with different capabilities.

An enhanced BCSM describes the behavior of the *Call Control Function* (CCF) operating in the B-SSP. The BCSM models calls that in contrast to the conventional IN are now initiated and controlled by the B-SCP ('third party call set up'). To account for the fact that an IN service will be related to more than one BCSMs, each corresponding to a single call, the concept of 'session' is introduced (Venieris, 1998). The session models the situation where a number of different calls and connections need to be co-ordinated so that a complex service is provided. At the B-SSP level the session is used as the context in which different BCSMs are correlated by the IN Switching Manager. This correlation is represented by the *Switching State Model* (SSM). The IN Switching Manager passes this information to the B-SCF residing in the B-SCP via appropriate B-INAP messages. To accommodate both the requirements for transparency of the network switching and transmission capabilities to the SCP, and the ability to describe complex network resource configurations, the SSM should adopt a flexible design philosophy. The developments presented in Venieris (1998) employ object oriented modeling of the SSM. Following these ideas, the session becomes an object that represents a complex call configuration. Other objects are the 'party' representing the user terminal or a network component (virtual party), the 'connection' which reflects the abstraction of a connection between the parties of a session, and the 'leg' object which represents the path to a party that is connected to other parties by a connection. Similar modeling approaches have been followed in the specification of advanced signaling systems (see, for example, Minzer (1991)).

Apart from the enhancements of the IN functional architecture towards a broadband environment, one should not overlook the developments around the signaling protocols of the B-IN. A prime objective of the B-IN concept is to minimize the modifications on standardized broadband signaling protocols. However, several additional features require a number of modifications and supplementary procedures with the most important the support of SCP initiated calls, and the User Service Interaction feature. In standardized signaling protocols, calls are always initiated by the users. The support of SCP initiated calls requires modifications and additions to the standard signaling procedures at the B-SSP and also special procedures and information flows at the B-INAP. Although the USI capability can be supported by simply enveloping user-service information to existing protocols, the B-INAP needs a number of extensions, for example the operation that is used by the B-SCP to instruct the B-SSP to monitor for user information from a specific user.

## 7.3    Distributed Intelligent Broadband Network

Though being a flexible and efficient approach to offering new services at low cost within the telecommunication environment, the Intelligent Broadband Network is a rigid architecture, where decisions about runtime service provisioning are taken within centralized B-SCPs. This has the drawback of limiting the overall number of calls handled by a network consisting of several switching systems and one B-SCP during a given time interval to the computing power of the B-SCP itself. This restriction is particularly important in the case of broadband services, where several calls take part in a service instance.

One of the most interesting aspects of the Intelligent Broadband Network resides in the clear distinction between the B-SCF, where service-related runtime control is maintained, and the B-SSF, responsible for handling call-related events and forwarding them to the B-SCF. Considering that in the Intelligent Broadband Network architecture the B-SSF is allocated to the B-SSPs and the B-SCF resides on the B-SCPs, this allows the definition of standard interfaces, based on the B-INAP protocol, between these distinct physical systems. By means of these interfaces, services can be developed and uploaded to the B-SCPs without modifying B-SSP software, provided that they comply with the definition of B-INAP operations. Benefits can be obtained by seeking distribution of intelligence within the network, while maintaining the advantages provided by the B-SSF/B-SCF separation.

An obvious growth in terms of computing power can be obtained by distributing B-SCF functionality within the network, bringing it close to the B-SSPs, and possibly integrating it within the same physical system. This increases the number of available B-SCFs and results in a higher number of calls handled by the IN architecture. The basic principle of separation between B-SSF and B-SCF is not altered.

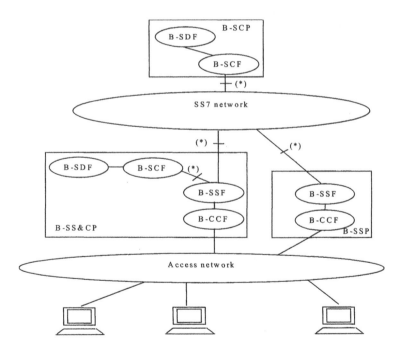

(*) = Broabdband INAP (B-INAP) protocol

**Figure 7.2** Distribution of B-SCF functionality

Figure 7.2 depicts a first example of enriched broadband IN environment. Enhanced switching systems, able to support IN services without the need for an external B-SCP, are named Broadband Service Switching and Control Points. These systems can coexist with traditional B-SSPs, and can communicate with B-SCPs by means of the B-INAP protocol. The presence of centralized Service Control Points is still useful, since not all services may be supported in a given instant within the B-SCF hosted by the B-SS&CP, depending on service allocation decisions made by network administrators. A straightforward choice may be to allocate frequently accessed services to distributed B-SCFs in B-SS&CPs, and maintain services with low access rate on centralized B-SCPs.

Traditional B-SSPs are still considered, because a full distribution of B-SCF functionality may not be realistic, due to economic considerations. A good balance between a fully centralized and a completely distributed approach can be obtained on the basis of the number, type and user acceptance of IN services within a geographic area administered by an operator.

The approach shown in Figure 7.3 provides a better distribution of network intelligence with respect to a fully centralized IN. However, location of this intelligence is not efficiently solved in the B-SS&CPs, which are based on the B-INAP protocol and on the underlying SS7 network. The SSF function contained in it has to look for services in the distributed B-SCF environment by means of a search scheme established *a priori* by network administrators. This is because SS7 has not been designed to comply with the principles of a *Distributed Processing Environment* (DPE), and therefore B-SSFs have to rely on a static algorithm searching for service intelligence.

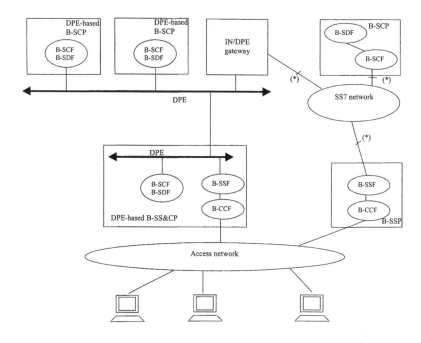

(*) = Broadbdband INAP (B-INAP) protocol

**Figure 7.3**  DPE-based B-SSF/B-SCF interface

The adoption of a standard DPE based on *Distributed Object Technology* can provide a promising way to overcome this drawback. Within this environment, functionality composing a distributed application can be allocated on distinct network nodes, and locator services provided by the DPE can be used to allow the localization of a particular component. Therefore, components do not need an *a priori* knowledge of the physical configuration of the distributed system.

The adoption of a DPE between B-SSF and B-SCF eliminates the need for static search criteria, therefore providing enhanced flexibility.

The approach depicted in Figure 7.3 provides great flexibility in localizing network intelligence. B-SSF and B-SCF are part of a distributed application relying on services provided by the DPE.

DPE communication substitutes an INAP-based one between B-SSF and B-SCF. However, the semantics for service handling are maintained, and the basic principle of B-SSF/B-SCF separation is still intact.

B-SCPs and B-SS&CPs based on DOT can coexist with traditional systems, which make use of special gateway functionality to communicate with DPE-enhanced systems. DOT-based B-SS&CPs can indifferently communicate with services available in their local SCF or available in remote DOT-based B-SCPs, by making use of DPE functionality (OMG IN/CORBA interworking, 1997a, 1997b).

Within DOT-based systems B-SCF and B-SDF FEs are shown together. Service Logic Programs and Service Data Tables for specific functionality will be included in the same object.

Though localization is now much easier, binding of network intelligence to the B-SCFs is still done on a static basis. Even greater flexibility is achieved by adopting an approach based on *Mobile Agent Technology* for the design of applications on the DPE. This allows services to be dynamically reallocated on the most appropriate B-SCF, depending on service specific needs.

A practical application is to start services in the centralized environment provided by a DOT-based B-SCP, and to move them closer to their users, i.e. to the B-SCF provided by the DOT-based B-SS&CP closer to the end systems, in case they are frequently accessed. This provides an efficient load control mechanism for B-SCPs, and allows service configuration to be tailored to actual usage.

In Figure 7.4 a DOT-based IN is used as a basis, and MAT is applied to the design and implementation of IN *Service Logic Programs* (SLPs). Services are started in a DPE and agent-based SCP, and can be moved to peripheral B-SS&CPs, depending on the service needs.

Initial configurations are designed offline by network administrators. This configuration may be later modified either by services themselves, which are provided with suitable decision algorithms, or by administrators as part of network management operations. With respect to the pure DPE-based solution, the advantage is that a change in the configuration is done in a more efficient way, being directly supported by MAT capabilities.

A DOT/MAT-based IN provides great flexibility, maintaining the basic distinction into Functional Entities and the related definition of interfaces which determined IN success. It is therefore reasonable to consider the integration of distributed, agent-based technologies as an important evolutionary step in the deployment of the Intelligent Network.

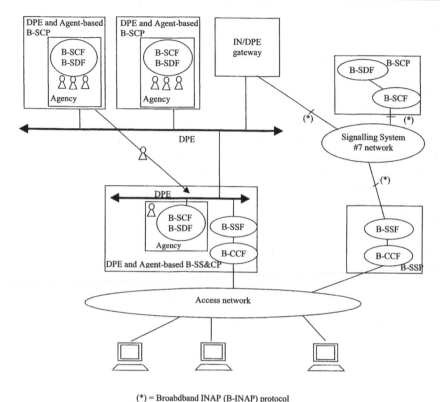

(*) = Broabdband INAP (B-INAP) protocol

**Figure 7.4**  MAT-based implementation of IN SLPs on top of DOT-based IN

## 7.4    Need for Interworking with Conventional IN/B-IN

Although DOT/MAT seems to provide a promising solution to problems typical of a centralized architecture, the introduction of these technologies in the Intelligent Network modifies the way IN systems interact.

As shown in the previous section, DPE-based B-SS&CPs and B-SCPs achieve communication via the DPE itself. This imposes the identification of an evolutionary strategy for the deployment of DOT/MAT-based IN (or distributed IN), which has to coexist with the traditional architecture, based on B-INAP protocols.

Distributed IN can be introduced in specific islands, which are interconnected to the main B-IN by means of adequate interworking functionality. The need for interworking is also shown in the previous section (Figure 7.3 and 7.4).

The interworking functionality is responsible for mapping B-INAP messages coming from the B-IN into DPE method invocations, and vice versa. Since B-INAP protocols are based on *Abstract Syntax Notation 1* (ASN.1), the gateway must handle syntax translation. This operation typically involves mapping ASN.1 data types into DPE-specific ones, and vice versa. The gateway also handles synchronization between B-IN dialogues and DPE method invocations.

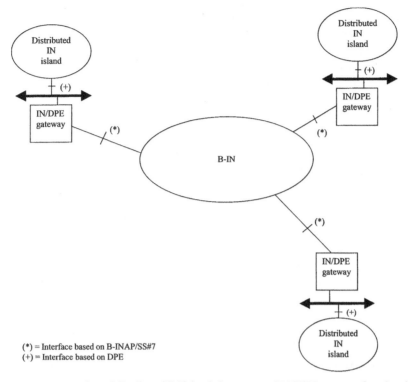

**Figure 7.5**   Interconnection of distributed IN islands by means of DPE/IN gateway functionality

## 7.5    Overview of Part III

The *Distributed Intelligent Broadband Network* (DIBN), as seen in Section 7.3, is an evolution of the Intelligent Broadband Network that enhances the flexibility of the service provisioning environment while increasing its programmability and maintainability. To do so, it makes heavy use of the most recent object oriented software technologies like DOT and MAT. However, it is still a quite new trend in telecommunications industry, and it hasn't taken the path towards the standardization or the market yet. Therefore, only prototype architectures and implementations of the DIBN currently exist. A study of such a prototype architecture, that has been developed within the framework of a European project (ACTS - MARINE), takes place in Chapters 8, 9 and 10 of this book. Through the analysis of the architecture, methodologies and real world components of this prototype DIBN system, one can watch how the advanced software practices, studied in the second part of this book, can be applied in systems that have been evolved within the telecommunications world in order to ameliorate their properties as well as their services. More specifically, Chapter 8 sketches the architectural properties of the DIBN and its differences compared to the centralized IBN. Chapter 9 provides an elaborate study of the advanced properties of the DIBN as these stem from the introduction of the DOT/MAT technologies in the architecture, at all levels of the system's procedures. These levels are the service provisioning, the service management, the service creation and the service switching level.

It also provides an overview of the technical properties of the system's physical elements. Finally, Chapter 10 prescribes three example services: the *Interactive Multimedia Retrieval* (IMR), the *Broadband Video Telephony* (BVT) and the *User Mobility Services*, that have been implemented in the prototype DIBN, and specifies by means of UML notation the dynamic behavior of the system during the execution of these services.

# References

Herzog, U. and Magedanz, T. (1998) Intelligent Networks and TINA – Migration and Interworking Issues, The Intelligent Network: Current Technologies, Applications, and Operations, pp. 109–122, ISBN: 0-933217-43-9, International Engineering Consortium (IEC), Chicago, IL.

Magedanz, T. and Popescu-Zeletin, R. (1996a) *Intelligent Networks – Basic Technology, Standards and Evolution*, International Thomson Computer Press, ISBN: 1-85032-293-7, London, June.

Magedanz, T., Rothermel, K. and Krause, S. (1996b) Intelligent Agents: An Emerging Technology for Next Generation Telecommunications? *Proceedings of IEEE INFOCOM '96*, pp. 464–472, IEEE Catalogue No. 96CB35887, ISBN: 0-8186-7292-7, IEEE Press.

Minzer, S.A. (1991) Signaling Protocol for Complex Multimedia Services. *IEEE Journal on Selected Areas in Communications*, **9**, 1383–1394.

OMG telecom/97-12-03 (1997a) Intelligent Networking with CORBA.

OMG telecom/97-12-06 (1997b) Interworking between CORBA and Intelligent Network Systems, Request for Proposal, December 8.

Venieris, I.S. and Hussmann, H. eds. (1998) *Intelligent Broadband Networks*. John Wiley & Sons Ltd.

# 8

# Architecture of the Distributed Intelligent Broadband Network

This chapter focuses on a description of the architectural principles of the *Distributed Intelligent Broadband Network* (DIBN). The general framework and expected benefits are considered in Section 8.1. Then the functional and physical architectures are defined in Section 8.2, in order to provide the basic reference material for the following part of this book. In the last part of the Chapter 9, more specific topics are dealt with. In Section 8.3 the methodology for moving from the high level functional architecture to the specification of the detailed components, and of their relationships, is analyzed, while in Section 8.4, the software architecture of some significant network elements is highlighted.

## 8.1    Introducing Advanced Software Technologies in the Distributed Intelligent Broadband Network

The Distributed Intelligent Broadband Network combines the technologies designed for a Telecommunication environment with the innovative methodologies and technologies available for the development of software applications that have been described in Part II.

A proper selection must be done from the different technical solutions in order to guarantee an open architecture, where the interoperability of the equipment provided by different manufacturers is an important requirement for the network operators. The two basic steps for moving from an *Intelligent Broadband Network* (IBN) to a Distributed IBN are the introduction of *Distributed Object Technology* (DOT) and *Mobile Agent Technology* (MAT).

Regarding the first step, the most promising Distributed Object Technology for a complex and rich architecture like DIBN is CORBA (see Chapter 4). In fact, CORBA is a standardized solution, and numerous commercial platforms are available in the Information Technology market. These platforms provide a set of CORBA services that enrich those features of the systems where the applications are developed and executed. Last but not least, the applications that exploit CORBA platforms can be written in different programming languages, and there is the possibility for reusing pre-existing legacy

*Object Oriented Software Technologies in Telecommunications*, Edited by I. Venieris, F. Zizza and T. Magedanz
© 2000 John Wiley & Sons Ltd

applications which do not use the object oriented approach, after an appropriate adaptation of the interfaces towards the other modules.

The control architecture of the DIBN is suitable to be considered as a distributed environment because both the users of signaling protocols (e.g. switch-based applications) and the users of IN protocols (added value telecommunications services) are placed in different network elements. Obviously, the possibility given by CORBA to replace the dialogue between users of Common Channel Signaling System Number 7 (ITU-T Q.700, 1993) based protocols with a client-server dialogue making use of *Internet Inter-ORB Protocol* (IIOP) (see Section 4.2.1.2) based protocols must be exploited, only if really beneficial. In Section 8.2 it will be indicated to what extent DOT is introduced in DIBN.

Moving to the second step, the basic requirement for the choice of a platform for development and execution of Mobile Agents is to be interoperable as much as possible with other platforms, and therefore to be *Mobile Agent System Interoperability Facility* (MASIF) compliant (see Chapter 6) (OMG MASIF, 1998).

Considering two or more MASIF compliant Mobile Agents Platforms, then, the choice is driven by other factors, like price and/or performance and/or additional services offered to the developers.

In Distributed IBN, Mobile Agents provide new opportunities in the areas of service management and service control in general. Services can be modeled as mobile service agents, which can be dynamically deployed in the network on the basis of the agent's intelligence. The extended availability of programmable systems in the network enables a decentralization of service intelligence.

Obviously, a synergy between the two steps is essential for the efficient employment of these advanced software technologies within Distributed IN. Agent platforms must be based on the chosen *Distributed Processing Environment* (DPE) in order to obtain a flexible and transparent (local or remote) service handling, according to service requirements. The basic advantages that are expected by the DIBN, if compared with IN and B-IN are:

- *Scalability*: the problem of having centralized points where Service Logic Programs are executed is solved by distributing the service code across the network when needed, thus being able to satisfy a higher demand.
- *Faster handling of service requests*: once the Service Logic Programs are downloaded closer to the users, a reduction of traffic on the signaling network occurs, thus leading to a better utilization of the network itself and to a consequent better service provision.
- *Dynamic service provisioning*: if a service is able to be executed both locally or remotely, it is possible to set criteria for deciding how the service must be provided; it is also possible to execute the same service locally in some areas and remotely in other areas of the same network.

Moving to the services that can exploit the DIBN architecture, it is possible to distinguish between two basic categories: downloadable/not downloadable services.

The services that belong to the first category can be executed indifferently locally (this is equivalent to have a distributed approach) or remotely (this is equivalent to have a centralized approach), while the services that belong to the second category must be executed only remotely.

The crucial point for identifying which category a given service belongs to is the analysis of the data associated with the service logic instance. If the data are mostly related to the user(s) involved in that particular service instance, then the service is downloadable. If the data is common to all the service logic instances concurrently running and cannot be duplicated for consistency problems, then the service is not downloadable.

The multimedia services described in Chapter 10 belong to the first category, while the mobility management procedures (described in Chapter 10 as well) belong to the second category.

## 8.2    The Distributed Intelligent Broadband Network Reference Architecture

The design of the DIBN architecture is not a trivial task. First, a merging between terms, models, representations, or in other words, between the 'language' of Intelligent Network and the 'language' of Object Oriented Information Technology is needed. In principle, two different approaches can be seen:

- To see IN as one of the domains of applications of Distributed Object and Mobile Agent Technologies.
- To consider these technologies as an enrichment of the classical IN architecture.

In the first case the dominant language will be that inherited from the Information Technology world, while in the second case the dominant language will be the IN one.

In this book the latter case has been preferred because it is considered important to identify the path from legacy IN systems towards the DIBN.

**Table 8.1**  Set of functional entities of the DIBN architecture

| Acronym | Name |
| --- | --- |
| B-CCF | Broadband Call Control Function |
| B-SCF | Broadband Service Control Function |
| B-SDF | Broadband Service Data Function |
| B-SRF | Broadband Specialized Resource Function |
| B-SSF | Broadband Service Switching Function |
| CCAF | Call Control Agent Function |
| M-CUSF | Mobility Call Unrelated Service Function |
| M-SCF | Mobility Service Control Function |
| M-SDF | Mobility Service Data Function |
| SCEF | Service Creation Environment Function |
| SCUAF | Service Control User Agent Function |
| SMAF | Service Management Access Function |
| SMF | Service Management Function |
| SCEF | Service Creation Environment Function |

According to this decision, the method of defining the architecture foresees the identification of the set of *Functional Entities* (FEs), the relationships between FEs and the mapping between the FEs and the *Physical Elements* (PEs).

In Table 8.1 the set of FEs present in the DIBN architecture is indicated in alphabetical order. Obviously, this set does not preclude enhancement for future needs.

The Functional Entities listed above have the following meaning:

- *B-CCF*. The Broadband Call Control Function is the functional entity that performs the following tasks:
  — Establishes, manipulates and releases broadband calls/connections as requested by the Call Control Agent Function
  — Provides trigger mechanisms to access IN functionality
  — Manages basic call resource data (e.g. call references)
  — Establishes, manipulates and releases broadband calls/connections as requested by a third party (e.g. this may be due to B-SSF requests on behalf of the SCP party).
- *B-SCF*. The Broadband Service Control Function is a function that commands Call Control functions in the processing of IN provided and/or custom service requests. It may interact with other FEs to access additional logic or to obtain information (service or user data) required for processing a call/service logic instance. In other words, the prime function of the B-SCF is the execution of Service Logic provided in the form of Service Logic Programs (SLPs). The B-SCF:
  — Interfaces and interacts with the Broadband Service Switching Function (B-SSF), the Broadband Specialized Resource Function (B-SRF) and the Broadband Service Data Function (B-SDF) Functional Entities
  — Contains the logic and processing capability required to handle IN provided services.
- *B-SDF*. The Broadband Service Data Function contains customer and network data for real-time access by the B-SCF in the execution of an IN provided service
- *B-SRF*. The Broadband Specialized Resource Function provides specialized resources, which are required for the execution of IN provided broadband services, in particular multimedia user interaction. In general, B-SRF:
  — Interfaces and interacts with a B-SCF
  — Is managed, updated and/or otherwise administered by an SMF
  — Contains the logic and processing capability to work together with broadband CPE in provisioning the services in a user-friendly way
  — May contain the communication and processing capabilities to make full use of the broadband transmission capabilities of the network.
- *B-SSF*. The Broadband Service Switching Function performs the following tasks:
  — Extends the logic of the B-CCF to include recognition of service control triggers and to interact with the B-SCF
  — Offers to the B-SCF an object view of the network resources and of the connection topology that compose the service instance
  — Modifies call/connection processing functions (in the B-CCF) by IN requests under the control of the B-SCF
  — Initiates call/connection processing functions (in the B-CCF) to process IN commands under the control of the B-SCF.

- *CCAF*. The Call Control Agent Function, as the agent between the user and the network call control functions, provides the following functionality:
  — Provides user access, interacting with the user to establish, maintain, modify and release, as required, a call or instance of service
  — Accesses the service providing capabilities of the CCF, using service requests (i.e. set-up transfer, hold, etc.) for the establishment, manipulation and release of a call or instance of service
  — Receives indications relating to the call or service from the CCF and relays them to the user as required.
- *M-CUSF*. The Mobility Call Unrelated Service Function is the Functional Entity that performs the following tasks:
  — Establishes, manages and releases the relationship between the instance in the SCUAF and the network for the call unrelated interaction between terminals and mobility management processing
  — Recognizes a call unrelated wired or wireless interaction to access IN and provides trigger mechanisms to access to the M-SCF
  — Modifies call-unrelated interaction processing functions to process requests for mobility management under the control of the M-SCF.
- *M-SCF*. The Mobility Service Control Function is a function that contains the overall mobility control logic. Service logic on the M-SCF is invoked by requests from other Functional Entities to support location and mobility management. The M-SCF includes the following functionality:
  — Location registration/de-registration
  — User registration/de-registration
  — Remote de-registration
  — User profile management
  — Paging control (e.g. initiation of paging, processing of paging response)
  — Local authentication processing
  — Confidentiality control (e.g. cipher management).
- *M-SDF*. The Mobility Service Data Function handles storage and accesses to service related and network data from the M-SCF, from the point of view of mobility management. It also provides consistency checks on the data.
- *SCUAF*. The Service Control User Agent Function provides call unrelated access for mobile users and/or wireless terminals. It supports the mobility management specific procedures not related to a service invocation. In general, the SCUAF includes the following functionality:
  — Initiation of call unrelated procedures for mobility management
  — Receives call unrelated service indications coming from the M-CUSF and relays them to the user/terminal as required
  — It supports procedures for registration/de-registration of mobile user/wired or wireless terminal in/from the network.
- *SMF/SMAF/SCEF*. The Service Management Function supports service introduction, provision, and maintenance and is accessed by a Service Management Access Function, providing the human-machine interface to the SMF. An additional Service Creation Environment Function allows for the specification, testing, and the introduction of services in the IN.

As a subsequent step, it is needed to identify which types of relationships (i.e. exchange of Information Flows) exist between the FEs. There are four possible types:

- *Call/connection control*: it is used to establish/modify/release the resources needed by the service using call control signaling
- *Call unrelated control*: it is used to establish a User to Service Interaction using call unrelated signaling
- *IN service control*: it is used to control the invocation/execution/termination of a Service Logic exchanging IN operations transported inside TCAP dialogues
- *Service Management*: it is used to create/deploy/manage the Service Logic Programs.

Table 8.2 contains the set of Relationships, the Functional Entities that use each Relationship and the classification according to the four types defined above.

It can be seen that the interfaces B-SCF <-> B-SDF and M-SCF <-> M-SDF are not present, because the object oriented approach requires a combination between the procedures making use of data and the data themselves.

**Table 8.2**   Set of functional relationships of the DIBN architecture

| Acronym | Endpoints of FEs | Type of relationship |
|---------|------------------|----------------------|
| R1  | CCAF <-> CCF | Call/Connection Control |
| R2  | SCUAF <-> M-CUSF | Call Unrelated Control |
| R3  | B-CCF <-> B-SRF | Call/Connection Control |
| R4  | B-CCF <-> B-CCF | Call/Connection Control |
| R5  | B-SSF <-> B-SCF | IN Service Control |
| R6  | B-SCF/B-SDF <-> B-SSF | IN Service Control |
| R7  | B-SCF/B-SDF <-> B-SRF | IN Service Control |
| R8  | M-SCF/M-SDF <-> M-CUSF | IN Service Control |
| R9  | M-SCF/M-SDF <-> B-SCF/B-SDF | IN Service Control |
| R10 | SMF <-> M-SCF/M-SDF | Service Management |
| R11 | SMF <-> B-SCF/B-SDF | Service Management |
| R12 | SMAF <-> SMF | Service Management |
| R13 | SMF <->SCEF | Service Management |

To progress the definition of the architecture, we need to identify where the Distributed Object Technology and the Mobile Agent Technology apply. From a theoretical point of view, every Functional Entity could contain CORBA objects and Mobile Agents. However, considering that the main objective of DIBN is related to an efficient service creation, deployment, distribution and execution, this means introducing DOT and MAT only in a subset of the Functional Entities, and consequently to using CORBA communication only for a subset of the Relationships.

A Functional Entity is defined as 'DOT based' if some (not necessarily all!) state machines/programs performing the required functions are implemented as CORBA objects, i.e. they are placed inside a CORBA Distributed Environment and offer an IDL interface to other client objects.

A Functional Entity is defined as 'DOT/MAT based' if some of the CORBA objects it contains have the capability of migrating, and are therefore implemented as Mobile Agents.

A Relationship is named 'DPE Relationship' if the transported communication flows are realized by means of the CORBA client/server paradigm.

A Relationship is named 'Not DPE Relationship' if the transported communication flows do not use the CORBA client/server paradigm, but utilize the usual signaling protocols.

Given these definitions, it is possible now to move to the overall Functional Architecture depicted in Figure 8.1.

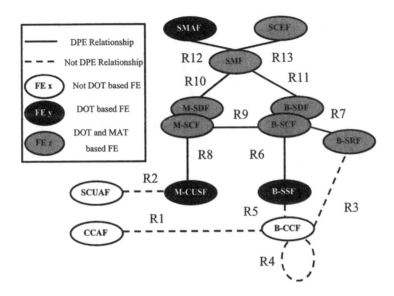

**Figure 8.1** DIBN functional architecture

It can be seen how the Functional Entities devoted to service creation (SCEF), service management and deployment (SMF) and service execution (B-SCF/B-SDF and M-SCF/M-SDF) are DOT/MAT based in order to guarantee the migration of Service Logic Programs.

The Functional Entities (M-CUSF, B-SSF) devoted on one side to detecting service invocations for interrogating IN, and on the other side to mapping IN instructions towards call/connection control commands are DOT based because they belong to the Distributed Processing Environment, but do not contain objects that need to migrate.

The Functional Entities dealing with user and network signaling for call/connection control do not belong to the DPE, and are therefore unchanged with respect to the B-IN architecture.

The typology of each Relationship is assigned in a straightforward way: service related interfaces are DPE Relationships; the interfaces related to the control of network resources are not DPE Relationships.

The following step defines the DIBN Physical Architecture, mapping the Functional Entities in the appropriate Physical Elements.

It is possible to observe that the IN distinction between FE and PE adapts well for the description of the DIBN because it allows us to represent the presence of Mobile Agents in different Physical Elements by means of the duplication of the FE where they are placed.

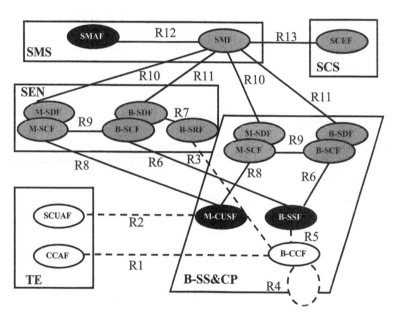

**Figure 8.2**  DIBN physical architecture

In Figure 8.2 the five Physical Elements of the DIBN are illustrated:

- *Terminal Equipment* (TE): the terminals are unaffected by the introduction of the Distributed Processing Environment; they invoke services and establish related resources by means of the same interface present in a B-IN network.
- *Broadband Service Switching and Control Point* (B-SS&CP): the B-SS&CP is an extension of a Broadband Service Switching Point with the capability to locally execute services downloaded from the network. Therefore, this network element is able to host both the B-SCF/B-SDF and the M-SCF/M-SDF Functional Entities. It is in principle possible for this node to also host the B-SRF, but it has been preferred to keep the Intelligent Peripheral functionality external to the B-SS&CP for keeping the R3 interface unaltered.
- *Service Execution Node* (SEN): this node joins the roles of the Broadband Service Control Point and of the Broadband Intelligent Peripheral. The SEN is able to receive the Service Logic Programs from the SCS and to download them to the B-SS&CP.
- *Service Management System* (SMS): this network element receives from the SCS the Service Logic Programs. SLPs are stored and moved to the appropriate network elements (SEN and/or B-SS&CP) both on a configuration and request basis.

- *Service Creation System* (SCS): it creates the Service Logic Programs by means of the residing Service Creation Environment. The availability of an agency within the SCS allows local service testing before the downloading of the SLPs to the SMS.

From the description of the network elements, it is clear that all the Physical Elements but the Terminals must contain an Agent Execution Environment. This is the most evident modification required to the network Physical Elements for being inserted into the DIBN Architecture.

## 8.3 Extending IN Design Methodology for the DIBN

In this section we examine how the design methodology of the classical Intelligent Network architecture should be augmented to accommodate the most significant new characteristics of the DIBN that are imposed by the adoption of DOT and MAT technologies. More specifically:

- In Section 8.3.1 a straightforward methodology for designing DIBN elements in an analogous way with the classical IN is illustrated.
- In Section 8.3.2 the issues that govern the implementation of the Functional Entity relationships in DIBN are identified.
- In Section 8.3.3 it is argued that the flexibility of a design methodology for the DIBN is completely aligned with that of classical IN.

### 8.3.1 Designing the functional entities

The classical methodology for designing an IN or B-IN Functional Entity consists of the following steps:

1. Decompose the functionality performed by the FE, in a set of tasks.
2. Assign to each task a Functional Block.
3. Specify inside each Functional Block the components, i.e. identify and define the state machines, data repository, protocol handlers that are needed to execute the task assigned to the Functional Block.
4. Identify and specify which primitives must be exchanged between the components.
5. Identify and specify which primitives must be exchanged between the Functional Blocks.
6. Identify and specify which messages must be exchanged between the FE and the other FEs when a Relationship exists.

The first phase is operated by the responsible standardization bodies. More specifically, ITU-T has defined the relationships between FEs placed in separate Physical Elements (external Interfaces originating a communication protocol), while the Relationships between FEs located in the same Physical Element (internal Interfaces not necessarily originating a communication protocol) are left to the implementers. Therefore manufacturers, when realizing their systems, proceed to the next steps.

Moving to the DIBN architecture, the decomposition of the system's functionality into Functional Entities remains as it is. However, the deployment of these entities into the Physical Elements of the system is more flexible, and can even be performed in a dynamic way. Furthermore, the communication mechanism is more flexible since the relationships are implemented by means of a DPE environment (see Section 8.2).

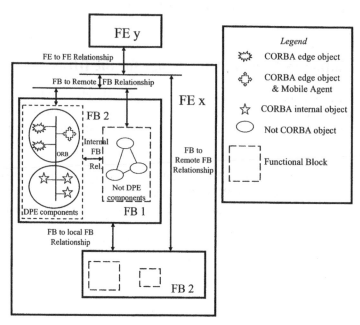

**Figure 8.3**   Decomposition of a functional entity

So the classical IN design methodology steps can be adapted as follows:

1. Decompose the functionality performed by the FE, in a set of tasks.
2. Assign to each task a functional block.
3. Identify inside each functional block the components belonging to the Distributed Environment with respect to the components not belonging to the Distributed Environment.
4. Specify inside each functional block the components not belonging to the Distributed Environment, i.e. identify and define the state machines, data repository, protocol handlers that are needed to execute the task assigned to the functional block.
5. Specify inside each functional block the components belonging to the Distributed Environment, i.e. identify and define the CORBA objects that act like a client or server.
6. Select which CORBA objects are internal objects, that means they do not interact with CORBA objects belonging to another Functional Entity, and which CORBA objects are edge objects, so named because they must interact with CORBA object belonging to another FE.
7. Select which CORBA edge objects may have the capability of migrating and therefore must be realized as Mobile Agents.

8. Identify and specify which primitives are exchanged between the Functional Blocks.
9. Identify and specify which messages are exchanged between the FE and the other FEs when a Relationship exists.

Figure 8.3 is a representation of the decomposition of a generic Functional Entity according to the methodology indicated above. We can see the decomposition of the Functional Entity into Functional Blocks and the assignment of the DOT/MAT characterization to those blocks. One can notice here that this approach of distinguishing the objects and characterizing the DOT/MAT properties of them is driven mostly by the classical decomposition of the IN architecture into functional components. This is contradictory with the object oriented principles for designing a system, but it is the most straightforward way to convert the concentrated IN architecture into a distributed one.

### 8.3.2    Implementing the functional entities relationships

As it is clear from the previous description about the design of Functional Entities, every relationship in the DIBN comprises information flows that are generated/transmitted by a set of components. These information flows can be realized in the following alternative ways:

1. When the two Functional Entities are both DPE based, the information flows between them can be transferred as IIOP messages through the CORBA bus that connects the two entities.
2. There is also the possibility that some of the communicating components in the DPE FEs are mobile agents. In this case, one or more of these agents can migrate to the physical place where its communicating peer resides. Such an implementation decision could either be advantageous or reduce radically the efficiency of the system.
3. If one of the two communicating Functional Entities is not DPE based, then the information flows among them are realized as TCAP messages (that is the transport layer of the SS7 protocol stack). In this case, if the other Functional Entity is DPE based, then it must have access to a gateway component that converts the IIOP messages to TCAP ones.
4. There may exist non-DPE based components in a DPE based Functional Entity. In this case, the non-DPE based components can be delegated by 'wrapper' objects (CORBA objects). Therefore, the information flows of this Functional Entity can be realized as IIOP messages.

In an IN domain where all the elements are DIBN enabled, cases 1, 2 and 4 can be applied. Case 3 is enforced whenever legacy IN elements interoperate with DIBN ones. Case 4 is enforced when a DIBN Functional Entity is built on top of legacy IN components or components implemented without a CORBA interface.

### 8.3.3    A more flexible approach in the DIBN design methodology

The methodology illustrated in Section 8.3.1 is completely aligned with the design philosophy of the classical IN. However, it is arguable that such an approach could unleash the flexibility and development productivity enhancement that DOT and MAT can offer. Nevertheless, one could adopt this methodology in order to stick to the traditional IN design process. In Chapters 9 and 10, where we examine the architectural details of a prototype DIBN implementation, a less strict methodology was followed that is more relative to the principles of distributed computing. In that prototype architecture the decomposition of the IN system into Functional Entities was preserved in order to produce DIBN components that interoperate with the legacy ones. However, the Functional Entities themselves were developed by means of object oriented practices. Moreover, in Chapter 9 we illustrate how the issues identified in Section 8.3.2 were elaborated to build an efficient communication backbone for the prototype DIBN system. This was based on CORBA and MAT to such an extent that a great portion of DOT and MAT facilities was exploited, and the drawbacks of the burdensome transactions of CORBA and MAT were minimized.

## 8.4    The Physical Elements

The DIBN architecture has defined new physical elements. In this section attention is paid to the Broadband Service Switching and Control Point (see Section 8.4.1), to the Service Execution Node (see Section 8.4.2), and to the Service Management System (see Section 8.4.3).

### 8.4.1    The Broadband Service and Switching Control Point

The B-SS&CP is designed enriching the *Broadband Service Switching Point* (B-SSP) of the B-IN architecture.

The B-SS&CP is physically composed by two subsystems: a *switching ATM matrix* and a *signaling server* for realizing signaling and IN tasks. The signaling channels are forwarded to the signaling server using semi-permanent connections established between the external interfaces to the other equipment and the ATM interface for accessing the signaling server. The two subsystems communicate by means of an Ethernet bus that allows the server to establish, modify and release switched connections, according to the requests contained in the signaling and/or IN messages.

Hereafter an overview of the evolution of the signaling server architecture for supporting the B-SS&CP capabilities is given. First, the B-SSP system is enriched in the following way:

- Introduction of DPE CORBA environment.
- Introduction of a Java Virtual Machine.
- Introduction of Grasshopper as runtime platform for mobile agents.

An adaptation of the existing software modules must be done: the DIBN architecture allows us to minimize the impact on the call related and unrelated signaling protocols, while the IN part (B-SSF and M-CUSF) needs major changes. Figure 8.4 shows the resulting B-SS&CP software architecture.

The signaling and Intelligent Network protocols (indicated in the figure with the boxes 'Access Signaling', 'Network Signaling', '*IN Application Protocol* (INAP) over *Signaling System 7* (SS#7)', 'Call-unrelated signaling') are not affected by the listed enrichments.

The *Switching State Model* (SSM) that manages the information about the network resources used by the different service instances is modified in order to be able to inter-operate with the DPE. The same applies to the *Mobility Basic Call Unrelated State Machine* (M-BCUSM) that is devoted to the detection of triggers/events for Mobility purposes, like User Registration.

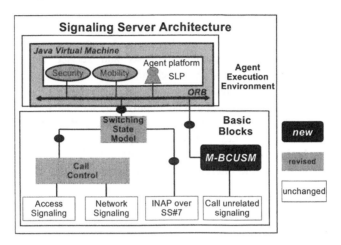

**Figure 8.4** B-SS&CP signaling server

## 8.4.2   The Service Execution Node

Services can run in a dedicated entity, named the Service Execution Node, which provides the capabilities needed for the user interaction and co-operation with the B-SS&CP. In terms of the conventional IN, this is the combined role of a *Service Control Point* (SCP) and *Intelligent Peripheral* (IP). Furthermore, the SEN integrates the introduction of DOT and MAT that are playing a key role in service deployment and execution within the Distributed IN (Prezerakos, 1998).

The SEN initially enables the B-SS&CP to access the IN agent-based services and provides an environment to locally execute the stationary or mobile agents that compose the services if this can result in performance gains. These agents are downloaded from the Service Management System, which constitutes the agent repository, during the service deployment. During actual service provisioning they can be downloaded from the SEN, according to service or network specific criteria. Essentially, the SLP operating in the SEN is cloned and the clone is instructed to migrate locally to the switch.

Alternatively, the SMS has the option of directly invoking the migration irrespectively of the specified criteria. This is necessary for instance if the SEN needs to be shut down for maintenance or in anticipation of a probable failure. Thus, several possible service execution scenarios can take place aiming at an optimized use of the network resources.

The SEN comprises the Broadband Service Control and Data Function, the Mobility Service Control and Data Function, the Broadband Specialized Resources Function, and the

Call Control Agent Function. The two major issues that arise with the introduction of DOT and MAT are the efficient provision of communication mechanisms among the physical entities, and the migration criteria of various software entities within the network.

- *Communication*. The SEN establishes a signaling connection with the Terminal Equipment to provide the user with the specialized resources required for the execution of a specific broadband service. The communication between the SEN and the B-SS&CP is based on CORBA, specifically IIOP, or on the SS7 signaling stack in order to support backwards compatibility with legacy SSPs. In the latter case, a B-INAP to CORBA bridge or gateway is used. Finally, the SEN communicates with the SMS via IIOP in order to serve management requests and service deployment, i.e. the downloading of agents from the SMS to their initial position within the network, which is of course likely to change subject to the specified migration criteria.
- *Migration*. The Service Logic Programs are deployed as stationary or mobile agents, and the latter can be distributed and executed among different physical entities. Several B-SS&CPs, some of which may generate a high number of service requests and consequently a high processing and communication load, can concurrently access a service in a SEN. A possible migration of the appropriate agents can improve network performance.

### 8.4.2.1    SEN's object model

The object model of the SEN considers five major parts. The first part is the essential software infrastructure that contains a commercial ORB and a Mobile Agent Platform for the support of MAT and DOT, respectively. The second part contains the components for the interaction of SEN with the other physical entities, namely the B-SS&CP and the SMS. The third part contains the core of the Service Logic, which is mainly the SLPs. The fourth part contains the objects responsible for the user interaction replacing the role of the Intelligent Peripheral in conventional systems. Finally, the IN-to-CORBA gateway constitutes the last part, which is necessary for interworking of the Distributed, CORBA-based IN systems with existing INAP-based systems. Figure 8.5 shows this high level decomposition of the SEN architecture.

The *Access Managers* are the SEN's gateway objects towards the B-SS&CP. There are two Access Managers: the B-SSF Access Manager, which is crossed by B-SSF/B-SCF interactions; and the M-BCUSM Access Manager, which is crossed by M-CUSF/M-SCF interactions. For simplicity, in the following only the B-SSF Access Manager will be referred to, but equivalent considerations apply to the M-BCUSM Access Manager as well.

The INAP information flows have been mapped into equivalent method invocations. Two CORBA servers (one for each direction) compose the B-SSF Access Manager; the servers are implemented using the multithreading approach so that multiple simultaneous connections between B-SSF and B-SCF objects can be established. The B-SSF Access Manager can be accessed either directly by the B-SSF objects, in case a CORBA-augmented INAP is used, or by the INAP-CORBA gateway, which in turn communicates with traditional switches using INAP. Since the B-SSF Access Manager is the only B-SSF visible entity, the gateway needs to interface with it only.

The *Service Logic Manager* (SLM) is the core entity for both B-SCF/B-SDF and M-SCF/M-SDF. It is responsible for locating the SLPs for each particular service that is requested, and for initiating migration procedures either triggered by the SMS or by the migration criteria that are applied. The SLM is, however, not aware of any service-level details of the SLPs it manages. When a service request arrives the SLM inspects the service key and examines its local data to get a CORBA reference to an SLP providing that service.

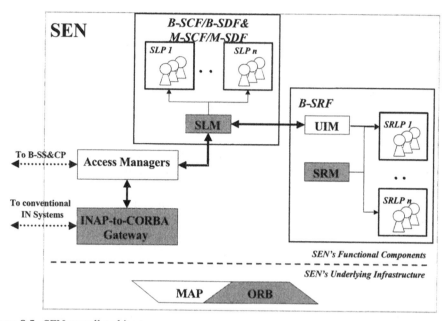

**Figure 8.5** SEN overall architecture

The *User Interaction Manager* (UIM) controls the lifecycle of the *Resource Logic Programs* (RLPs) and acts as the front-end of the SRF entity by publishing the SRF functionality in the CORBA environment of the DIBN. The RLPs are DOT/MAT components that control the user interaction phase of the Terminals with the Specialized Resources within the service sessions.

The network provider, on behalf of different service providers, can exploit the capabilities of the B-SRF. For example, if a service provider wishes to make new content available he will inform the SMS about the name of the content and its descriptive attributes (such as the location of the content server or the category that this new content should appear to user terminals). An agent containing this information will be created at the SMS and is then instructed to migrate to the B-SRF agency. Using the Region Registry's facilities, the agent will attach to the *Specialized Resource Manager* (SRM). After this attachment, new users connected to the B-SRF and browsing through its catalogues will be able to find the new content available.

Exactly like the Service Logic Manager and the B-SSF Access Manager are the SLPs' 'connectivity' mediators to the B-SSF (and vice versa); the User Interaction Manager is the SLPs' mediator to the B-SRF. The interface between the SLPs and the UIM is also based on CORBA. In this way, all SLP references to their environment, as well as the environment's references to the SLPs, are DPE-based allowing location transparent communication and enabling a potential migration. The SRF hosts a number of Resource Logic Programs that are implemented as MAs in the same manner as are the SLPs. The

RLPs need access to underlying transport facilities for exchanging data with the user terminals during the user-service interaction phase.

### 8.4.3   The Service Management System

The Service Management System is responsible for the lifecycle management of the services which are realized in terms of a set of interacting, distributed mobile agents. The following lifecycle phases are handled by the SMS:

- *Deployment phase*: creation and initial customization of a service instance and transfer of this instance into the *Environment of Service Provisioning* (ESP).
- *Utilization phase*: monitoring and control of service instance during its runtime within the ESP.
- *Maintenance phase*: replacement of a running service instance.
- *Withdrawal phase*: removal of a service instance from the ESP.

Concerning the management point of view, the SMS environment is divided into two parts: the SMS agency, representing the access point for administrators; and a set of ESP agencies, i.e. agencies residing within the environment of service provisioning and hosting *Service Agents* (SAs) during their runtime (see Figure 8.6). The SMS itself is realized as a set of different management agents:

- The *Service Management Controller* (SMC): acts as a management access point for administrators and resides inside the SMS agency. Its main task is the co-ordination of all management activities and the maintenance of management-related data, such as runtime information about all active service instances.
- The *Messaging Service Agent* (MSA): supports the exchange of messages between SMS components and service instances. If agents acting as message senders or receivers migrate to another location, the MSA is responsible for re-directing all future messages corresponding to the sender's/receiver's new location. (The messaging service corresponds to notification/event mechanisms in traditional management systems.)
- The *Managed Object Agents* (MOAs): are distributed across the ESP and provide local management access to the service agents. The SMC may delegate certain management tasks to a MOA in order to unburden itself. According to its policy, an MOA either remains at a specific ESP agency or follows a specific service agent through the environment.
- The *Rule Agents* (RAs): are used to provide managing capabilities to MOAs.

To achieve distributed managing functionality, a *management context* may be specified for each MOA, defining its autonomous managing behavior and allowing an administrator to delegate management tasks to the MOA. For instance, a management context may define the operation(s) to be performed due to the retrieval of specific events. A possible management context could be the following: 'if the notification A, raised by supplier B, has been received N times within M seconds, then perform the management operation C on service agent D'.

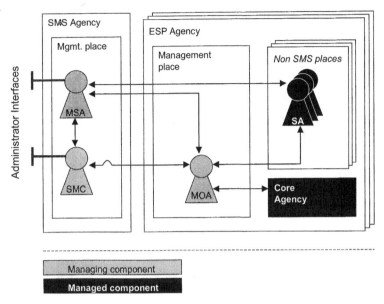

**Figure 8.6** Overall SMS architecture

Management contexts are realized by a set of *Rule Agents* (RA). A Rule Agent is a mobile agent whose core part specifies a certain event-driven managing behavior. The task of a rule agent is to retrieve specific events that occur in the hosting agency, to analyze them, and to react according to these events, e.g. by performing certain actions. To be notified about events, a Rule Agent registers itself at the local Managed Object Agent. Along with this registration, the RA specifies a set of trigger events. If the MOA (which is by default connected with the hosting agency and permanently listening for agency-related events) determines the occurrence of a specific event, it browses its trigger table. If there is any registered RA interested in the event that occured, the MOA informs this RA, which in turn reacts according to its internal policy. Figure 8.7 shows the sequence of steps that are followed in the Mobile Agent based management by delegation approach.

1. The administrator wants to delegate certain autonomous managing behavior to an ESP agency. Thus, the administrator contacts the SMC to create a new rule agent that provides the intended behavior.
2. The administrator customizes the new rule agent via the RA's user interface. Among others, the intended destination agency has to be specified.
3. The rule agent migrates to the ESP agency.
4. The rule agent registers itself at the local MOA. Along with this registration, the RA informs the MOA about the events it is interested in. The MOA creates a new entry in its trigger table, specifying that the new RA has to be contacted after the occurrence of the events previously delivered by the RA.
5. An event occurs in the agency (such as the creation, migration, or termination of an agent).
6. The MOA browses its trigger table and determines that a registered RA is interested in the event that occurred.
7. The MOA informs the RA about the event.

8. The RA analyzes the event according to its internal policy and determines how to react.
9. The RA reacts to the event by performing an operation on a real resource.

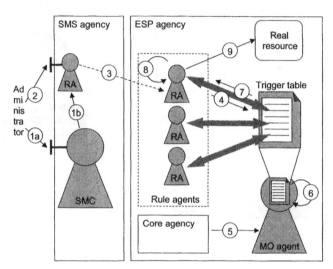

**Figure 8.7** Mobile agent based management by delegation

# References

ITU-T, Recommendation Q.700 (1993) Introduction to CCITT Signaling System No. 7.

OMG, MASIF-RTF Results (1998) http://www.omg.org/docs/orbos/98-03-09.pdf .

Prezerakos, G., Pyrovolakis, O. and Venieris I. (1998) Service Node Architectures Incorporating Mobile Agent Technology and CORBA in Broadband IN. *Proceedings of IEEE/IFIP MMNS'98*, Versailles, France, November.

# 9

# Deployment of DOT/MAT Technology into the Distributed Intelligent Broadband Network

Nearing the end of the twentieth century, the *Intelligent Network* (IN) has become the most important technological basis for the rapid and uniform provision of advanced telecommunication services above an increasing number of different transmission networks. In line with the main objectives of the IN, namely to provide a network and service independent architecture, the IN can be regarded as a middleware platform for a telecommunications environment, primarily based on the incorporation of state of the art information technology into the telecommunications world.

One fundamental principle of the IN, and probably the key to its success, is its inherent evolution capability, also expressed by the notion of standardized IN *Capability Sets* (CS). This phased approach of standardization and thus evolution allows us, on the one hand, to extend the IN service capabilities and architecture as required by the market demands, i.e. by extending the scope of IN service provision from *Public Switched Telephone Network* (PSTN) and *Integrated Digital Services Network* (ISDN) into *Broadband ISDN* (B-ISDN), *Public Land Mobile Network* (PLMNs), and most recently the *Internet* and *Voice over IP* (VoIP) networks. On the other hand, IN evolution takes into account the technological progress in the information technology world. Whereas the first IN architectures are based on exploitation of the concepts from the distributed computing world, such as the remote procedure call paradigm, and the availability of common channel signaling networks in the seventies, newer IN architectures are driven by the emerging *Distributed Object Technologies* (DOT) and *Mobile Agent Technologies* (MAT) in the nineties (Magedanz, 1996; Herzog, 1998).

This chapter discusses the evolution of the Intelligent Network towards a Distributed IN, analyzing how the network itself is currently structured, and identifying principles enabling migration towards the distributed IN. The important intermediate step of Intelligent Broadband Networks, which defines new resource models suitable for the control of advanced services, is also outlined.

*Object Oriented Software Technologies in Telecommunications*, Edited by I. Venieris, F. Zizza and T. Magedanz
© 2000 John Wiley & Sons Ltd

## 9.1    What CORBA Offers to the DIBN Architecture

As discussed in Chapter 7, Distributed Object Technology provides a good basis for the deployment of a *Distributed Intelligent Broadband Network* (DIBN), due to its application transparent handling of resource location within a network. Among the possible implementations of DOT, the *Common Object Request Broker Architecture* (CORBA) standard presented in Chapter 4 represents a very promising solution.

CORBA provides developers with a series of advanced features, used to build object oriented distributed applications. In the DIBN context, the most interesting features are as follows:

- Location transparency
- Standardized interfaces between modules (objects)
- Dynamic Invocation Interface (DII)
- Naming Service (NS).

### 9.1.1    Location transparency

A major issue in developing the DIBN is abstraction from the actual location of functionality within the network. Due to the nature of the Distributed Intelligent Broadband Network, service requests may be accomplished by centralized servers as well as by service intelligence close to the switching system.

Therefore, the independence from the actual location of service objects and the ability to dynamically locate them are key issues in providing services in the distributed environment.

The locator services provided by CORBA fulfil this task. CORBA clients can rely on the *Object Request Broker* (ORB) to locate the server able to supply the requested service (OMG/CORBA, 1998). In the DIBN scenario, this feature is used when the *Broadband Service Switching Function* (B-SSF) requests a service from the Broadband *Service Control Function* (B-SCF). This *Functional Entity* (FE) may be either located in a centralized *Broadband Service Control Point* (B-SCP) or within the same system hosting the B-SSF, the *Broadband Service Switching and Control Point* (B-SS&CP) (see Chapter 7).

Object localization can be improved by using a naming service, as briefly presented later in this chapter.

### 9.1.2    Standardized interfaces between modules

The CORBA standard considers the definition of clear interfaces offered by server objects to client ones as one of its strongest points.

The specification includes an *Interface Definition Language* (IDL, see Chapter 4), by means of which server interfaces are defined as a set of attributes specific of the server objects and operations (methods) allowed on it. Once an object has been specified in terms of its IDL interface, developers can work independently on the client and the server side on the basis of this interface definition. This implies that no internal details of the server object are expected to be necessary to the client developer (apart from an obvious information about its expected behavior).

A set of rules for mapping IDL to conventional programming languages (C++, Java, etc.) is part of the CORBA specification. This allows clients and servers to be developed by means of distinct languages if needed.

In the DIBN environment, IN *Service Logic Programs* (SLPs) act as CORBA servers, while B-SSF functionality plays the role of CORBA client. IDL here basically applies to the definition of interfaces offered by SLPs residing in the B-SCF Functional Entity to client objects residing in the B-SSF FE. Thanks to locator services and to the Naming Service, the location of these objects in the distributed environment is not relevant.

### 9.1.3 Dynamic Invocation Interface

While a basic IDL definition foresees that the interface offered by a server object is statically known to its client(s), there may be cases in which the interface is not known *a priori*. CORBA solves this issue by providing the *Dynamic Invocation Interface* mechanism.

The DII allows a client to have a dynamical view of interfaces provided by servers it wants to contact. This is made possible by the provision of an Interface Repository, which is browsed by clients in order to retrieve details of the servers' interfaces. In the DIBN case, this feature may be useful to avoid a static definition of interfaces provided by SLPs. This would allow the definition of new SLPs independently from a statically defined framework.

### 9.1.4 Naming Service

The *Naming Service* (NS) is an additional CORBA component (i.e. it is not part of the core CORBA services, see Chapter 4) which allows client objects to retrieve *Interoperable Object References* (IORs) of server objects from a database common to all objects within a given domain. An IOR univocally identifies a server object in a domain.

Since the IOR also contains the address of the network node on which the server object runs, it may be regarded, to a certain extent, as a complement to locator services.

Using the NS has the advantage of eliminating the need for a runtime binding of clients to servers. However, due to faulty situations, the NS may contain old references (as an example, the server object stopped executing due to an exception and the IOR is still in the NS). Therefore, these exceptions must be considered in the design of an application using the NS.

Despite its flexibility advantages, another drawback of using the NS is that its centralized nature within a domain implies that its failure may hinder the behavior of the whole domain.

Used within the DIBN scenario, the NS allows a greater flexibility in the allocation of resources to different physical systems.

## 9.2 The Communication Backbone

Figure 9.1 shows the architecture of a prototype DIBN system. Overall, DIBN is characterized by: broadband, *Asynchronous Transfer Mode* (ATM) bearer services and the use of a distributed infrastructure (CORBA and MAT) for the IN-control part (see Section

7.3). The most important change in the infrastructure of the DIBN compared to that of a traditional B-IN is the substitution of the message-based *Transaction Capabilities Protocol* (TCAP) by a *Distributed Processing Environment* (DPE) as the means to carry the IN flows between FEs. Note that this substitution does not change the semantics at higher layer protocols such as the *Intelligent Network Application Protocol* (INAP). As seen from previous sections, the basic elements of the DIBN architecture are analogous to those of the traditional IN. The DIBN architecture is enhanced at the level of component implementation technology and communication infrastructure. Therefore, even if the transport mechanisms are more sophisticated than the traditional higher layer protocols of IN, the semantics of the communication dialog among the Functional Entities of the system remain unchanged. Hence, the service agents of the DIBN communicate with interfaces whose semantic information is the same as in INAP protocol.

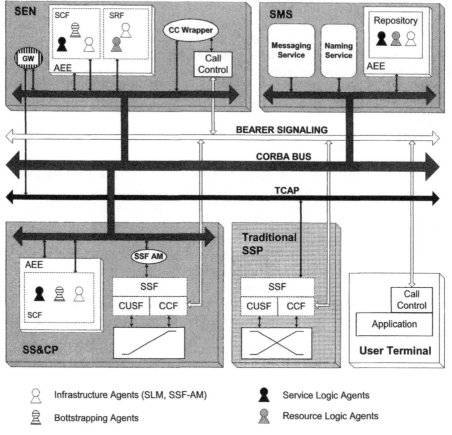

**Figure 9.1**   High level physical view of the DIBN

The architecture depicted in Figure 9.1 differs from that adopted in the traditional B-IN in the following ways:

1. *Service Logic Programs* (now called *Service Logic Agents*) are autonomous software components (mobile agent) that are deployed dynamically and maintain direct CORBA interfaces with INAP semantics with their peers.
2. There are not only interfaces between Functional Entities, but there are also direct interfaces between components that belong to different FEs. This is facilitated by CORBA technology and promotes the distributed design of the system.
3. Specialized Resources are controlled by *Resource Logic Agents* created within the *Service Creation Framework*. Hence, the SRF is extensible and can be updated in a completely dynamic manner.

These differences, in spite of being of great significance, do not alter the basic concepts of the *Intelligent Network Conceptual Model* (INCM). On the other hand, they constitute the basic advantage of DIBN in comparison to the traditional IN.

The grouping into SCF and SRF objects depicted in Figure 9.1 reflects a logical discrimination of components according to their functionality rather than defining black boxes like the Functional Entities of the traditional IN. This design approach emanates from the principles of the *Distributed Processing Environment*, where the components are executed in a virtual homogeneous environment that is spread across multiple physical places.

Ultimately, the resulting DIBN system is completely interoperable with traditional B-IN systems, and even further, it can embed traditional B-IN elements within its own domain. In the system illustrated in Figure 9.1, a traditional B-SSP is depicted. As we can see, the only difference between the SSF residing in the SS&CP (MAT and DOT enhanced) and the SSF that resides in the traditional B-SSP is that the former uses the CORBA bus as a communication medium, while the latter uses the TCAP. Both SSFs can trigger services in the SEN through two different paths:

- The SS&CP SSF through the SSF Access Manager that resides in the SS&CP, which in turn contacts the *Service Logic Manager* (SLM) that also resides in the SS&CP.
- The B-SSP SSF through the Gateway that resides in the SEN, which in turn contacts the SLM that resides in the SEN.

The only translation functionality needed is that implemented in the Gateway. However, even this functionality is very simple, since it is all about mapping one-to-one the INAP operation coming as TCAP messages to INAP method calls through CORBA, and vice versa.

Overall, the DIBN retains compatibility with the centralized B-IN at both system and element level. The latter is very important because it can facilitate the step-by-step introduction of the DIBN into existing centralized system. For instance, you don't have to upgrade all the SSPs in an existing IN domain to SS&CPs if you need to introduce a MAT based *Service Execution Node* (SEN). Old SSPs can interwork with the DIBN SEN and yet exploit the advanced SEN features. This facility can open the way for the seamless introduction of Distributed B-IN technology in real world Intelligent Networks.

## 9.3    Exploiting MAT Migration Facilities in the DIBN

*Mobile Agent Technology* (MAT) (see Chapter 6) is used in the DIBN architecture to enhance the flexibility of the service provisioning system, as well as to improve non-functional parameters of the system like operational flexibility and specific performance characteristics. The most significant facility offered by MAT is agent migration (see Section 6.3.1.5). By utilizing MAT migration facilities in the DIBN, it is possible to implement dynamic (in a sense) code allocation and reallocation scenarios, optimizing the system's configuration even during its operation. This section describes the infrastructure used in the DIBN that provides the necessary substrate to implement the aforementioned scenarios, and analyzes alternative schemes that can be applied to exploit agent migration facilities in an effective manner.

### 9.3.1    Applying MAT in the DIBN

In Chapter 7 we saw that the DIBN system is formed by the interconnection of three types of network elements; the *Service Management System* (SMS), the *Service Execution Node* and the *Service Switching and Control Point*. We have seen also how these elements correspond to the physical elements of a traditional B-IN architecture. In this section we will see how each of these elements can be enhanced by employing the Mobile Agent Technology.

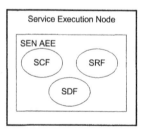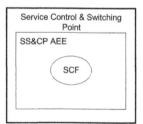

**Figure 9.2**  Types of Agent Execution Environments in the DIBN

The network elements of the DIBN among other components comprise an *Agent Execution Environment* (AEE). The AEE is a computational runtime environment that can receive, store, manage and execute specific kinds of agents that control and manage the provision of IN services. The AEE is actually a Java-based *Mobile Agent Platform* (MAP) (see Section 6.3) that is augmented with service management facilities. The AEE plays a specific role in the system, depending on the network element where it is hosted. There are three types of AEE (Figure 9.2):

- *SMS AEE*: this AEE resides in the SMS. The SMS AEE serves as a repository for all the mobile agents of the system, and also hosts the management service environment. In terms of an IN, this AEE comprises the Service Management Agent Function.
- *SEN AEE*: this AEE resides in the SEN. The SEN AEE controls and executes all the agents that relate to the execution of an IN service session. In terms of an IN, this AEE

comprises the Service Control Function, the Service Data Function and the Specialized Resources Function.

- *SS&CP AEE*: this AEE resides in the SS&CP. The SS&CP AEE controls and executes those agents and objects that relate to the control logic which is needed for the execution of an IN service session. In terms of an IN, this AEE comprises the Service Control Function.

These AEEs can host various types of agents. In the DIBN architecture we can divide the agents into the following categories:

- *Service Logic Agents* (SLA): these are the agents that incorporate the Service Logic Program of an IN service (see Section 7.1). Each agent class type corresponds to a specific IN service type. Whenever a new IN service session is initiated, its corresponding agent is created and activated in order to control the session. The Service Logic Agents resided originally in the *Service Management Agent Function* (SMAF) (Venieris, 1998) place, and during service provisioning at the SCF places.
- *Resource Logic Agents* (RLA): these are the agents that control the specialized resources provided to a service user within an IN service session. Each agent is responsible for controlling the resources within a specific session. The resource logic agents resided originally in the SMAF place and during service provisioning at the *Specialized Resource Function* (SRF) places.
- *Infrastructure Agents* (IA): the infrastructure agents implement objects whose main role is either to control or to provide wrapping functionality to Service and Resource Logic Agents. The Infrastructure Agents have two parts: a stationary part and a mobile one. The mobile part contains information that is subject to change, and therefore may need to be updated during the operation time of the DIBN system. The stationary part comprises only the basic mechanisms related to the control of other agents, and therefore need not be updated. The mobile part resides in the SMAF, SCF and SRF, while the Stationary in SCF and SRF.
- *Bootstrapping Agents* (BA): the bootstrapping agents create all the stationary parts of the Infrastructure Agents. There exists one bootstrapping agent for each place in the AEEs.

The SMS AEE acts as the repository of all the agents that are able to migrate (Mobile Agents). These are all except for the stationary part of the Infrastructure Agents (which are Stationary Agents). Just before the initialization of the system, all the Mobile Agents reside only in the SMS AEE while the Stationary Agents reside in the AEE where they will operate. During initialization the following migrations take place (Figure 9.3):

- The Mobile Parts of the *Edge Agents* (EAs) migrate from the SMS AEE to their relevant AEE where their corresponding Stationary Parts are.
- The Resource Logic Agents, one per IN service type, migrate from the SMS AEE to the SRF Place of the SEN AEE.
- The Service Logic Agents, one per IN service type, migrate from the SMS AEE to the SCF Place of the SEN AEE.

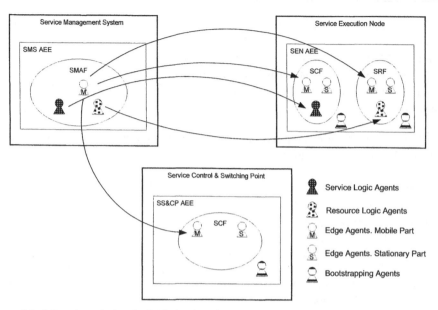

**Figure 9.3**  Migrations during the initialization phase

During the operational phase of the system, agents may be cloned and migrate among the AEEs. The following migrations and clonings may occur during this phase:

1. Cloning of a Service Logic Agent. This may happen due to the initiation of a new IN service session.
2. Migration of a Service Logic Agent from the SEN AEE to the SS&CP AEE. This may happen due to performance requirements, as we will see in the next paragraph.
3. Cloning of a Resource Logic Agent. This may happen due to the initiation of a new dialog among the SRF and a user during a specific IN service session.
4. Migration of new versions of an SLA, RLA or EA Mobile Part from the SMS AEE to the relevant AEE due to an update procedure (see Section 9.5).

As case 4 occurs quite infrequently, after relatively long time periods, we shall examine here case 2, while cases 1 and 3 will be analyzed more elaborately in subsequent sections.

As mentioned before, there are as many classes of Service Logic Agents as there are types of IN services. During the initialization phase, a single instance of every type is created (but not yet activated), and then it migrates to the SEN AEE. During the normal operation phase, whenever a new user in the system requests the initiation of a new IN service session, the relevant SCF Place is located in either an SEN or an SS&CP AEE, and the initial instance (*First Generation Prototype*, FGP) of the corresponding Service Logic Agent is cloned once. Depending on the state of the system, this instance may be activated and executed at the place where it was cloned (Worker), or it may migrate to an SS&CP AEE SCF place (*Second Generation Prototype*, SGP) and be cloned again there (Worker). The terms FGP, SGP and Worker are applied to the various instances of the same Service Logic Agent class type. The following constraints are considered:

- An FGP can be cloned to either a Worker or an SGP
- An SGP can be cloned only to a Worker
- A Worker cannot be cloned or migrate
- Only a Worker can be activated and executed
- When an SGP is cloned by an FGP it must migrate to a different SCF place that resides in an SS&CP AEE where there is no other SGP, instance of the same class type.

All the above constraints result in the following consequences:

- In each SEN AEE there is exactly one FGP for every Service Logic Agent class type
- In each SS&CP there may be at most one SGP for every Service Logic Agent class type
- Workers are clones of either an FGP or SGP, and are activated and executed immediately after their cloning and at the same place.

In Figure 9.4 the possible alternative cases for the cloning and/or migration of a new Service Logic Agent instance are depicted.

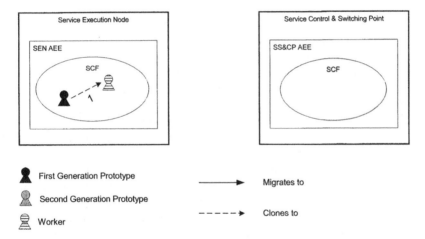

**Figure 9.4**   SLA instantiation case A; the worker is cloned directly from the FGP

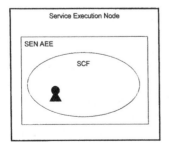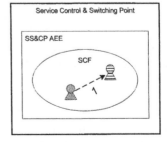

**Figure 9.5**   SLA instantiation case B; the worker is cloned directly from the SGP

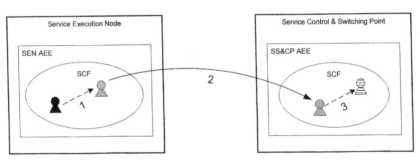

**Figure 9.6** SLA instantiation case C; the FGP clones an SGP, this migrates to an SS&CP AEE and clones there a worker

The instantiation cases illustrated depict the different ways that the DIBN agent-based service components can be created and allocated during the runtime operation of the system. In these cases, the service initialization and update operations are not included. These mainly involve the management procedure, and will be analyzed in Section 9.5.

### 9.3.2    Dynamic balancing of service control load

Service Logic Agents and Resource Logic Agents are those components in the DIBN architecture that perform the control of every active IN service session. We saw in the previous section that the initialization of a new IN service session results in the creation of a new SLA instance by cloning. RLA instances are also created for each session, but their creation depends upon the way that the service session execution evolves. This means that when the load in a DIBN system increases (more users request for an IN service), then the number of active Service Logic and Resource Logic agents in the system increases in a linear manner.

In a classical B-IN architecture, the SLPs that control the execution of every IN service session reside and execute in the Service Control Point (equivalent to the Service Execution Node of the DIBN. See Section 8.2). Therefore, the entire load that is produced by the service control processes is concentrated to the Service Control Points (Figure 9.7).

Typically, in real world IN architectures the SCPs control a large number of SSPs. Since SCPs are pure IN elements, they are burdened only with the IN service load of the system, which usually represents a small fraction of the total load of the network. In the special cases where the IN load may exceed the anticipated one, the SCP may become congested, and thus the responsiveness of the network will be reduced or, even worse, the SCP may become blocked. In the DIBN architecture there is the capability of transferring service control load from the SEN (equivalent to the IN SCP) to the SS&CP (equivalent to the IN SSP). This capability, if it is used with the proper criteria, can offer a very good solution not only to marginal situations like SCP congestion, but also for performance optimization.

Transferring of control load from SEN to SS&CP means that fewer Service Logic Agents will be executed on the SEN and more on the SS&CP. However, there are different approaches to realize this scenario, which appear at two levels. The first level includes the definition of the criteria that will be used to decide when the transfer takes place. The second level deals with the way in which and the number of agents that will be transferred.

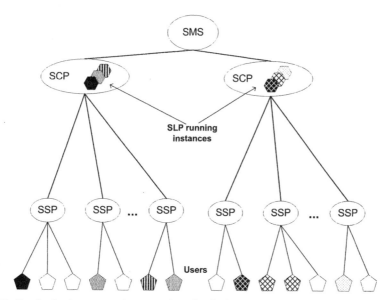

**Figure 9.7** Service logic program instances in a classical IN while a number of users have open IN service sessions

Starting from the second level, it is very crucial to take into consideration the network and processing overhead of MAT operations like migration and cloning. For instance, migrating every single SLA instance from SEN to SS&CP would produce more load even to the SEN than executing the instance there. On the other hand, cloning adds no network load and much less processing overhead than migration. So, a simple and effective scheme is to migrate only prototype SLAs from the SEN, if these do not exist in the destination SS&CP. There, whenever a new SLA instance is required, the prototype SLA will be cloned and the clone will be activated in order to control a new service session. This scheme can become clearer if you consider SLA instantiation cases A, B and C (see Figure 9.8). Specifically, this scheme is constituted by the following steps:

1. Initially, the first SLA instances of every IN service type are created at the SMS and are called *First Generation Prototypes* (FGP). All FGPs then migrate to the SENs, but they are not activated. When some users request various IN services from their SS&CPs, these requests cause the creation of several SLAs in the central SEN that controls the relevant SS&CPs. These creations are actually cloning operations of the FGPs that correspond to the requested IN service types (see Figure 9.8, instantiation case A). Each clone in that step is a Worker and is activated and executed in order to control a specific IN service session. At this step all IN service sessions are controlled by SLAs which reside in the SEN(s).

2. If the load in the SEN or the network connections increases too much, it may be more efficient to transfer some control load from the SEN to the SS&CPs. So, with the use of special performance criteria, some SS&CPs will stop relaying the spawning of new SLA clones to their responsible SEN and will start creating their own locally (see Figure 9.8, instantiation case B). However, initially there is no SLA prototype instance residing at the SS&CPs. Hence, a prototype instance should be instantiated from a corresponding one that resides in the SEN, and migrate to the SS&CP. This is called *Second*

*Generation Prototype* (SGP, see Figure 9.8, SLA instantiation case C). So, only prototype instances migrate, and only once per IN service type. In this way, the number of migrations in the system in minimized while the load transfer from the SEN to the SS&CP is achieved.

Considering now the criteria that judge the decision to activate step 2, it is evident that these are strongly correlated to the particularities of the specific IN network implementation. However, a simple approach would be to define thresholds of the number of activated SLA instances to the SEN and the SS&CPs. The thresholds in the SEN will force the processing load established there to always be under an acceptable upper limit. The thresholds in the SS&CPs in turn will regulate the load at their network connections with the SEN. These connections are loaded mostly by the number of CORBA transactions among the activated SLA instances and the SSF components. On the other hand, SS&CPs should not be overloaded because these elements are also burdened with the provisioning of bearer services to the users. Therefore, the performance criteria that affect the balancing of the service control load in the system should take into consideration all the performance parameters of the network elements.

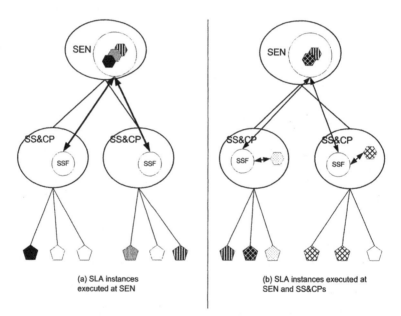

(a) SLA instances executed at SEN

(b) SLA instances executed at SEN and SS&CPs

**Figure 9.8**  SLA instances in the DIBN during the operation of several IN service sessions

Overall, the utilization of migration facilities in the runtime operation of the DIBN system provides a flexible infrastructure that allows the dynamic configuration of the distributed components according to the performance requirements of the network. This became evident from the service control load balancing capabilities that this infrastructure can offer, enabling the optimization of the network elements' efficiency.

## 9.3.3    Functional view of the DIBN

The architectural principles of the DIBN were described in Chapter 8. In this section a brief overview of the functional structure of the DIBN is given. Special focus has been put on the SCF and SRF entities in order to examine how the dynamic configuration schemes, seen in the previous sections, can be applied.

In the DIBN, SCF and SRF Functional Entities, whose role is prescribed in the IN standards, deviate from their traditional substance not in the roles they assume, but rather in the technology that is chosen to implement them and the different communication infrastructure they use. Both the SCF and the SRF are populated by MAs, which provide the logic for different aspects of their operation. Programmatic hooks are implemented that can be used by these objects to plug onto the CORBA bus and invoke methods on other agents or stationary objects, or have methods invoked on them. Because of the nature of the technologies used to implement SCF and SRF, these Functional Entities can be considered as an homogeneous environment where software components can be executed and communicate with each other regardless of their location; this is called a *Service Execution Environment* (SEE). SEE actually comprises the set of SCF and SRF Functional Entities that belong to the same *Service Provisioning Domain* (SPD). This set is ultimately constituted by the SCF and SRF located at a specific SEN, along with the SCF residing in the SEN connected to this SS&CP.

The architecture of the Service Execution Environment is depicted at a basic level in Figure 9.9. The objects that are grouped in the same Functional Entity do not necessarily reside at the same network place. For instance, in most cases the Service Logic Agents reside at the SEN while the *Service Logic Manager* that controls them resides at the SS&CP. However, these objects can be logically grouped in the same Functional Entity since their functionality is closely related with the SCF functionality of the traditional B-IN. In the following, the functionality of the main modules of the SCF and SRF groups is described:

- *Service Logic Manager* (SLM). This module is constituted by Infrastructure Agents (see Section 9.3.1), and has both a Stationary and Mobile part. It communicates via a CORBA interface with the SSF indirectly (through SSF Access Manager). Its main role is to control the lifecycle of the Service Logic Agents. When the SLM receives a request from the SSF for the creation of a new service session, it creates the corresponding Service Logic Agent either locally or remotely. Therefore, this module applies the performance criteria in order to perform balancing of the service load.
- *SSF Access Manager* (SSCF-AM). The SSF Access Manager acts as an intermediate between the SSF objects and the Service Logic Agents. It maintains CORBA interfaces with both SLAs and SSF objects, and relays the invocations among them. The SSFAM can connect the relevant SLA with its peer SSF object, keeping track of the active SLAs' network location. Therefore, the main purpose of its relaying functionality emanates from the desire to hide from the SSF objects the actual location of the SLAs.
- *User Interaction Manager* (UIM). This module controls the creation of the Resource Logic Agents. The RLAs are always created locally, wherever UIM resides. RLAs control resource related objects stored in the SEN. UIM has a mobile and a stationary part. The mobile part contains information that is subject to frequent changes, like the association of Service Keys with RLA types, or specific parameters for the creation of

RLAs. The stationary part implements the mechanisms for controlling the RLAs, and provides the interfaces among the RLAs, SLAs and SLM.

- *Specialized Resources Objects.* These objects handle the data of the specialized resources like web scripts, multimedia content or meta-data. They also implement server-side protocols for operating dialog with the user's application.
- *Terminating Call Control Agent Function* (CCAF). RLAs have to interface with the Call Control Function of the SEN in order to handle the establishment of user-plane connections between the Users' ATM terminals and the SRF Specialized Resources Objects. This module acts as a CORBA wrapper over the Call Control module of the SEN. It provides CORBA interfaces to the SLAs that enable them to correlate User-plane connections with the active IN sessions.
- *Data Connection Adapter* (DCA). This module provides wrapping functionality of the *ATM Adaptation Layer 5* (AAL5) services for sending and receiving packetized data through a user-plane connection. Its interfaces are used by the Specialized Resources Objects in order to exchange data with the user applications.

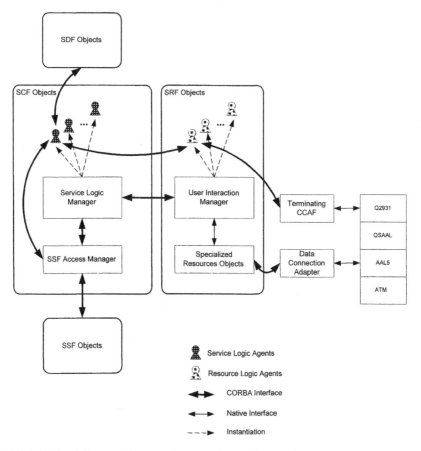

**Figure 9.9**  High level diagram of the service execution environment

## 9.4    Service Creation Methodology and Framework

Service creation in a traditional Broadband-IN system is an off-line procedure that develops new IN services. The development process of a new IN service does not affect elements or components of the existing IN system. New IN services can be introduced into the system by plugging in adjunct components that encompass the logic of the new services. These components are the Service Logic Programs. In many implementations of traditional B-IN systems the incorporation of new SLPs into the system requires the recompilation of part of the code in the Service Control Function. Even in cases where the recompilation is not required, the deployment of the new SLPs in the *Service Control Point* (SCP) demands the restarting and reconfiguration of the SCP, an inconvenient procedure that interrupts normal operation of the system.

In a Distributed IBN, the SLPs are incorporated into the Service Logic Agents. The SLPs represent the pure logic of the services that is embedded in the SLAs, while functionality for the dynamic deployment of the components in the system is also incorporated. Hence, the introduction of new IN services in the DIBN system is very flexible and dynamic, because the SLAs that convey the service logic can be deployed even during the operation time of the system in the service enabled elements of the system (see Section 9.3.2). Regarding the procedure of creating IN services in DIBN, we can examine an example. Consider an *Interactive Multimedia Retrieval* (IMR) service which involves several parties like a user terminal, the *Intelligent Peripheral* (IP) (embedded in the SEN, see Section 9.3.3) and one or more *Multimedia Servers* (MS). In a typical scenario of IMR, the user is first connected to the IP where he browses service-related information, and then the system connects him with his favorable MS. The IMR SLA monitors the execution state of this scenario and controls the network resources (creation and destruction of network links, Quality of Service selection, etc.) that are required for the service to be offered. Moreover, in IMR service a Resource Logic Agent controls the interaction of the user with the IP in co-operation with the SLA. To develop the logic of the IMR service, one has to follow a software lifecycle of specifying the requirements, designing the IMR SLA and RLA components, implementing them using Java, the CORBA platform and the Mobile Agent Platform, and testing it in the real system. In this case, he would have to tackle the intricacies of both the CORBA and MAP platforms and deal with non-service related issues like migration under specific circumstances, and cloning to generate new service instances. So, there is the flexibility and usability improvement in the DIBN if service creation becomes equivalent to a software development process, often complicated with special know-how requirements for the service designer. Legacy IN and B-IN systems usually encompassed powerful *Service Creation Environments* (SCE), where the service designer had only to pinpoint icons (pre-defined primitive components rarely corresponding to standardized SIBs), interconnecting them to each other and fill-in some dialog-based forms to parameterize them. This procedure was an easy one, and the work of the service designer was simplified. However, these systems were closed, not able to seamlessly embed user-written code in order to enhance the functionality of the services. Hence, the creation of more complex services was tricky and demanding.

In the DIBN object oriented software development practices can boost the creation of complex services. Moreover, the burdensome requirements that a build-from-scratch SLA or RLA task implies can be significantly alleviated with the use of a Special Framework that encompasses all functionality related to the intricacies of the MAP and CORBA platforms and migration and cloning operations. Therefore, only code implementing the core functionality of the service logic needs to be written in order to create a new service. Furthermore, parts of already developed services can be reused to create new ones, as object oriented techniques permit. For instance, to create a more complex information retrieval service, IMR SLA can be used as the basis, while IMR RLA can be reused unchanged.

## 9.4.1    Service creation framework

An example of such a framework is briefly described in this section. This framework was employed in a prototype DIBN system that was based on commercial MAP and CORBA platforms.

The class library provided by the MAP offers the base MA class (`MobileAgent`) which is extended to produce the actual SLAs and RLAs. However, the developer of the service is not immediately using the `MobileAgent` class. The architecture of the system mandates that certain capabilities must be built into the SLAs. For instance, the Service Logic Manager should be able to create workers and a SGP out of FGP or to instruct them to migrate or register themselves in the CORBA Naming Service. Clearly, programmatic hooks must be available on the SLPs to enable the SLM to perform these tasks. All the capabilities used by other elements are referred to as environment adaptation capabilities, and should clearly not concern the developer of an IN service. For this reason, service creation is based on a *Service Logic Framework* that extends MAP's class library further. In doing so, it provides specialized skeletal classes and interfaces that implement the functionality of SLAs related to their lifecycle, management and basic service control tasks. This framework deviates from the traditional SIB libraries one in that it adopts object oriented engineering practices. Also, since all environment specific provisions are built into the framework, it offers a neat way of guaranteeing that service developers can concentrate on the logic of their services and be oblivious to the environment in which they will operate. The service creation framework consists of two largely orthogonal class hierarchies:

- *Environment Adaptation Class Library* (EACL). EACL implements all the environment specific tasks of the SLAs like controlling cloning and migration operations and performing CORBA related operations like retrieving object references from the NS, connecting to the ORB or handling threaded servers. Moreover, EACL encompasses agent lifecycle functions and management related operations such as the removal or searching for an agent in the RR. Therefore, EACL constitutes a rather 'thick' layer for abstracting the intricacies of the underlying CORBA and MA platforms. It also caters for architectural decisions in a manner transparent to the developer. In the event that the CORBA or MA product used should change, all required modifications can be accommodated in the EACL requiring no changes in the IN service designer's code (or to the SIF – see below). This would also be the case if, for instance, the logic of the SLM would change to allow, e.g. for third generation prototypes or a different handling of the second generation ones. A developer is therefore not directly extending MAP's MobileAgent class, but rather EACL's `CompositeAgent` class, and need not

concern himself with CORBA servers, MA migration or worker-SGP creation. A high
level view of the EACL class library is depicted in Figure 9.10.

- *Service Intelligence Framework* (SIF). The SIF comprises base classes, skeleton code
  and pre-implemented design patterns to aid the service designer in encoding the actual
  intelligence for an IN service. More specifically, SIF provides *Switching State Machine*
  (SSM) classes, *Finite State Machine* classes and classes for automating the generation of
  the necessary INAP related messages. SIF's class diagram with some example classes
  for an Interactive Multimedia Retrieval service is depicted in Figure 9.11. Use of SIF,
  however, in contrast to the use of EACL, is optional. While it is a class library an IN
  service developer would find helpful – and in fact we have implemented an IMR SLA
  using it – one could just as well implement an IN service from scratch. Using the EACL,
  on the other hand, is mandatory. A developer has to extend the
  `CompositeServiceAgent` class. Not doing so would prevent an SLA from being
  used by the architecture.

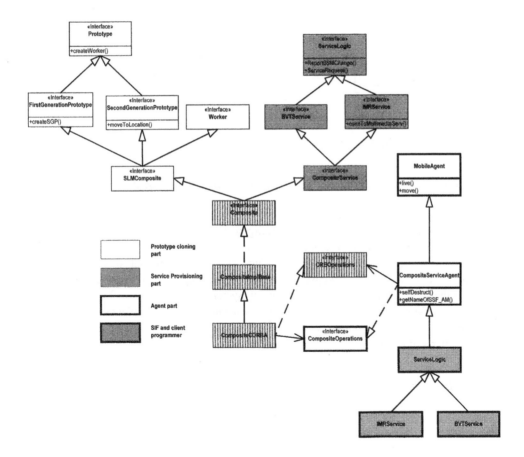

**Figure 9.10**   EACL class library high level view

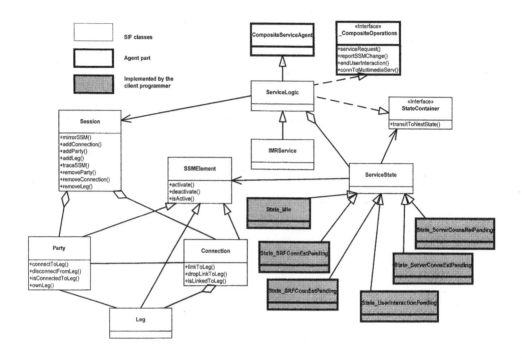

**Figure 9.11**   SIF class library high level view

## 9.5    Service Management Mechanisms and Procedures

The task of the Service Management System within MARINE is to monitor and control mobile agent-based Service Logic in an Intelligent Network environment. These services are composed of numerous interacting mobile agents.

The purpose of this section is to introduce such a system that was designed and developed within MARINE (Breugst, 1999). The remaining part of this section will first introduce the scope of the Service Management System. Thereafter, the following section describes the SMS architecture as an overview by means of agencies, places, and agents, in the following referred to as *components*. Also, a more detailed introduction to the core composed parts of the SMS is given.

### 9.5.1    Scope of Service Management System

Among others, the Service Management System is responsible for several phases of a service lifecycle, which are explained in the following paragraphs. Some of these phases cover service independent as well as service specific procedures. The performance of the service specific procedures, such as the data exchange between old and new service agent instances during a service update, have to be performed or at least supported by the service developers.

- *Service Deployment.*[1] This phase covers the functionality responsible for the introduction of a new service instance into the *Environment of Service Provisioning* (ESP). This phase comprises the creation, customization, deployment, and activation of a service instance as well as the configuration of the ESP.

- *Service Utilization.* This phase covers all management aspects defined by the different functional areas: *Fault, Configuration, Accounting, Performance,* and *Security* (FCAPS)[2], as well as access management aspects that are related to service instances during their execution. The utilization includes the monitoring, control, and operation of a service instance by providers as well as customers. The capability provided during this phase can be split into functions that are common to all or a specific service instance. The Service Management System provides different kinds of management agents to support these capabilities. Since agents carry out the management tasks, they can use the mobility facility provided by the underlying MA platform to deploy themselves closely to the service agents, thus taking advantage of local interactions. Due to their inherent autonomy, a manager can delegate a management context to the management agents in order to enable them to initiate management procedures and react automatically to notifications emitted by the service agents.

- *Service Maintenance.* This phase comprises the procedures required for the version management of a service instance, such as an update of a complete service instance or just parts of it during its runtime. This update is a complex task, covering the creation of new (updated) service agents, the retrieval of the current execution state from the corresponding (old) instances, the data exchange between the old and the new instances, and the replacement of the old ones by the new ones at the same locations. All occupied resources have to be released before an old service agent is removed, and they have to be occupied again by the new agent instances. These interactions require a careful synchronization between all entities involved.

- *Service Withdrawal.* This phase handles the removal of a service instance from the ESP without any negative impact on the environment or other service instances. The Service Management System triggers the withdrawal and announces it to all 'interested' entities by raising suitable notifications. However, each service agent is responsible for performing its own termination, e.g. by releasing resources.

### 9.5.2    Service Management System architecture

The Service Management System, designed and implemented within MARINE, is a distributed, component-based system realized by means of different management components. These components are either located within a designated SMS agency, serving as an access point for administrators, or within the agencies in which the service provision takes place, i.e. the Environment of Service Provisioning. The SMS is composed of the so-

---

[1] Note that two previous phases exist before the service deployment can start, i.e. the service design and service provisioning phases. These phases are out of the scope of the Service Management System, and it is considered that they are performed within a service creation environment in which services are designed, specified and implemented.

[2] FCAPS functional areas identify the functional capabilities that service management should consider. This acronym is a well known term of the OSI functional classification.

called Core SMS (see Section 9.5.2.1) which provides general, service independent management capabilities. This core part is a compound of four components: the *Service Management Controller* (SMC) (see Section 9.5.2.2), the *Managed Object Agent* (MOA) (see Section 9.5.2.3), the *Monitoring Agent* (see Section 9.5.2.4), and the *Messaging Service* (see Section 9.5.2.5).

In the following, after an overview, a more detailed insight into these components will be given.

### 9.5.2.1    Core Service Management System

The SMS is a system composed of those elements as identified in Figure 9.12. As shown, the core SMS consists of four agent-based components:

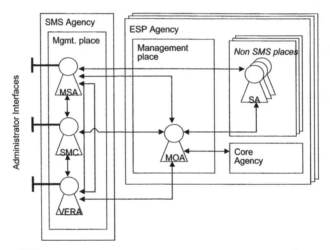

**Figure 9.12**   Core SMS

- The *Service Management Controller* is the component that controls, coordinates and maintains connections to the management agents within the ESP, and provides management access for an administrator by means of graphical user interfaces. Via these interfaces, the administrator is able to control the service lifecycles and the environment of service provisioning according to the SMS use case model. For instance, the installation, maintenance, and withdrawal of service instances is initiated by an administrator via the SMC.
- Each agency within the ESP hosts a *Managed Object* (MO) *agent*. This agent provides general management access to the core agency. Note that the MOAs provided by the SMS support only service independent management functionality. Thus, they are associated with an ESP agency rather than with specific service agents. However, if required, service-specific MOAs (S-MOAs) may be realized by service developers. These S-MOAs can be associated with a single service agent in order to realize service-specific management capabilities.
- The SMS agency hosts an *enVironmEnt monitoRing Agent* (VERA) which keeps track of the components (agents and agencies) within the ESP. The VERA is aware of the activities taking place in the ESP, and reports them to the administrator. For this

purpose, the VERA uses, depending on its configuration, information emitted by MOAs and/or the MSA.

- The SMS agency hosts a *Messaging Service Agent* (MSA), which is responsible for receiving and delivering notifications between components in the MARINE environment. Each agent may connect to the messaging service as consumer and/or supplier.

### 9.5.2.2    Service Management Controller
The Service Management Controller is the initial SMS access point for administrators. This agent provides graphical user interfaces in order to enable human actors to control the service lifecycles and the environment of service provisioning. For this purpose, the SMC has access to a repository, maintaining service templates and profiles, a service agent code base, and environment specific information.

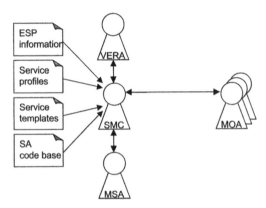

**Figure 9.13**   Service Management Controller

On the other hand, the SMC is connected to the environment of service provisioning via the messaging service as well as the Managed Object Agents that are residing in the ESP agencies.

To update its ESP-related information base, the SMC is connected to the environment monitoring agent. For instance, the SMC may contact VERA to search for a specific service instance during its runtime or to check the availability of ESP agencies.

### 9.5.2.3    Managed Object Agents
For the management of service agents, an approach similar to the OSI *Managed Object* (MO) approach is considered (ITU-T/X701, 1992; ITU-T/X720, 1993). The service agents represent real resources, i.e. the entities that have to be managed by the SMS. Similar to the OSI approach, all management operations are not directly performed on the real resources (i.e. the service agents), but instead on specific management agents, the Managed Object Agents. Each MOA is responsible for the management of at least one service agent. In contrast to OSI systems management, where a Managed Object is just a passive management interface to a real resource, an MOA additionally covers functionality

associated with an OSI agent[3], and even with an OSI manager. In particular, the realization of managing functionality within an MOA represents a powerful enhancement compared to the traditional, rather centralized management systems. Due to the MOA's inherent autonomy, specific management contexts can be defined by a manager and sent to an MOA to automate certain management tasks. In this way, the managing functionality can be decentralized across the distributed environment.

Figure 9.14 shows the relationship between the traditional and agent-based management approach.

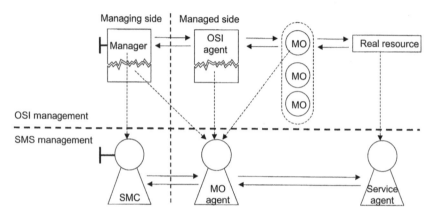

**Figure 9.14** OSI management versus SMS management

The access point at the managing side is realized by means of the Service Management Controller, which is responsible for the co-ordination of management activities and provides access to human actors (administrators).

Figure 9.15 shows the functionality of a MOA. The separation of the agent into a core, access, usage, and management part is derived from the USCM object structure defined in the *TINA Service Architecture* (TINA-C, 1997).

The actual task of an MOA is the management of a real resource which, for instance, can be a service agent or an agency. Thus, this managing functionality is defined in the MOA's core part. Its usage part, providing access to the core part, offers several interfaces to be accessed by external entities, particularly managers. The usage as well as the management part provides several interfaces for accessing the computational, management behavior of an MOA agent.

**Generic Management Interface**
The Generic Management Interface can be used to perform 'generic' management operations on real resources. These operations are not specific to the service agent to be managed, and enable among others its suspension, resumption, and migration, as well as

---

[3] In the context of OSI management, the term *agent* has a completely different meaning than in the context of mobile agent technology. To distinguish these terms, an agent in the context of OSI management is referred to as *OSI agent*. An OSI agent is an entity associated with an OSI managed system which builds the bridge between the managed objects and the managing system (i.e. the manager).

the retrieval of service independent attributes, such as the service agent's state. To perform these management tasks, the MOA accesses the core functionality of the underlying MA platform.

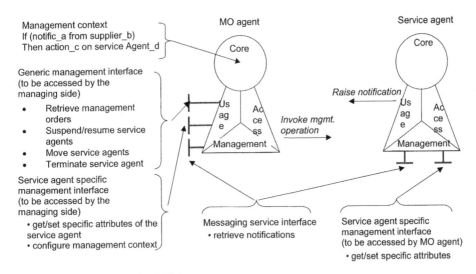

**Figure 9.15**   Functionality of an MOA

**Messaging Service Interface**
The Messaging Service Interface allows a MOA to retrieve notifications that are raised by any other component running in the distributed environment. The messaging service is an integrated part of the SMS (see Section 9.5.2.5). It supports a MOA to monitor the behavior of the managed resource, and to react in a suitable way, e.g. by performing specific operations due to its internal management context. Besides, a MOA itself may raise notifications to be retrieved by interested consumers.

**MOA specific management interface**
The MOAs initially provided by the SMS (called *default MOAs* in the following) realize service independent management capabilities. The associated real resource is actually the hosting ESP agency rather than a specific service agent. To invoke management operations on agents residing in the local agency, the MOA accesses the agency's core interface. That means the default MOAs are not able to perform service specific operations on service agents, such as the modification or retrieval of specific attribute values, since they do not provide any SA-specific interfaces. If such enhanced functionality is required, the developers of service agents may implement a service specific MOA (S-MOA) that is associated with a single service agent.

**S-MOA specific management interface**
In addition to the functionality of a default MOA, an S-MOA may provide a service agent Specific Management Interface that enables a manager to perform operations and to access attributes that are specific to the managed service agent. This specific part of a MOA is similar to the access part of a managed object in the traditional OSI approach. Regarding

OSI management, the developer of the real resource specifies this part in terms of *Guidelines for the Definition of Managed Objects* (GDMO) definitions (ITU-T/X.722, 1992). These definitions are used by a manager to generate managed objects. In the scope of MARINE, the service specific management functionality must be specified and implemented by the service developer and added to the default MOA. That means a service developer may enhance a default MOA with service specific management functionality.

Note that, to be manageable, each service agent must provide management interfaces accessible by the corresponding MOA. On the one hand, this covers service specific operations to be invoked and attribute values to be read or modified, and on the other hand, notifications that may be raised by the MOA. These interfaces are associated with the management part of the service agent. An initial management interface is inherently provided by means of the agent's super classes (i.e. the classes `Service`, `MobileAgent`, `StationaryAgent`). If additional interfaces are required, they have to be realized by the service developer.

### Management Context

To realize distributed managing functionality, a *management context* may be specified for each MOA, defining its autonomous *managing behavior*, and allowing an administrator to delegate management tasks to the MOA. For instance, a management context may define the operation(s) to be performed due to the retrieval of specific events. A possible management context could be the following: 'if notification A, raised by supplier B, has been received N times within M seconds, then perform the management operation C on service agent D'. Regarding S-MOAs, no suggestions or restrictions are made concerning the management context, since this is an implementation matter, handled by the respective developer. Regarding the default MOAs, management contexts are realized by a set of *Rule Agents* (RAs).

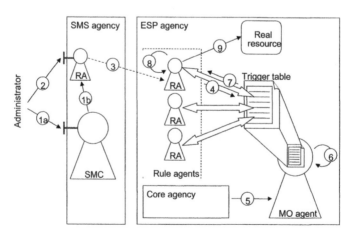

**Figure 9.16**  Management context via rule agents

A Rule Agent is a mobile agent whose core part specifies a certain, event-driven managing behavior. The task of a rule agent is to retrieve specific events that occur in the hosting agency, to analyze them, and to react according to these events, e.g. by performing certain actions. To be notified about events, a rule agent registers itself at the local Managed Object

Agent. Along with this registration, the RA specifies a set of trigger events. If the MOA (which is by default connected with the hosting agency and permanently listening for agency-related events) determines the occurrence of a specific event, it browses its trigger table. If there is any RA registered that is interested in the event that occurred, the MOA informs this RA, which in turn reacts according to its internal policy. Figure 9.16 shows this approach.

- *Step 1*. The administrator wants to delegate certain autonomous managing behavior to an ESP agency. Thus, the administrator contacts the SMC in order to create a new rule agent that provides the intended behavior.
- *Step 2*. The administrator customizes the new rule agent via the RA's user interface. Among others, the intended destination agency has to be specified.
- *Step 3*. The rule agent migrates to the ESP agency.
- *Step 4*. The rule agent registers itself at the local MOA. Along with this registration, the RA informs the MOA about the events it is interested in. The MOA creates a new entry in its trigger table, specifying that the new RA has to be contacted after the occurrence of the events previously delivered by the RA.
- *Step 5*. An event occurs in the agency (such as the creation, migration, or termination of an agent).
- *Step 6*. The MOA browses its trigger table and determines that a registered RA is interested in the event.
- *Step 7*. The MOA informs the RA about the event.
- *Step 8*. The RA analyses the event according to its internal policy and determines how to react.
- *Step 9*. The RA reacts to the event by performing an operation on a real resource.

The approach described allows a very flexible configuration of the managing behavior of an MOA, since its management context can be modified simply by adding, changing, or removing the associated rule agents. Any context logic may be realized whose complexity is just limited by the power of the programming language used (i.e. Java in the MARINE context).

### 9.5.2.4    The environment monitoring agent

The task of the monitoring agent is to track activities within the ESP, to log them, and to display them to the administrator. The VERA is connected to the environment and uses the Grasshopper platform facilities for its monitoring tasks.

The VERA provides a user interface that allows the administrator to get a view of the ESP. Via policies, the administrator is able to specify the monitoring, logging, and display behavior. Apart from the user interface, the VERA provides an API to be used by other SMS components, such as the Service Management Controller. Among others, the SMC contacts the VERA to retrieve environment-related information that is required for the deployment, maintenance, and withdrawal of a service instance.

### 9.5.2.5    The Messaging Service

The *Messaging Service* (MSS) is represented by an agent, the so-called *Messaging Service Agent*, which resides on the management place of the SMS agency. The SMS agency serves as an access point to the management capabilities for administrators for fulfilling their management tasks. The management place is a particular location within the agency that provides multiple management capabilities to objects acting in a management role. The MSA provides the notification forwarding mechanism within the distributed environment of MARINE. This environment forms a domain of notification exchange, the so-called *Notification Forwarding Domain* (NFD). For the sake of simplicity, a focus on the inter-domain behavior of the messaging service is not considered.[4] Notifications are only forwarded within such a domain.

Associated with such an NFD are different types of agents and agencies, acting in different roles:

- *Supplier agent*: an object that uses the messaging service to send notifications via an associated channel.
- *Consumer agent*: an object that receives notifications from the messaging service via an attached channel.
- *Channel*: the logical communication link between a supplier agent and a set of consumer agents via the messaging service. All channels within a NFD are controlled by the MSA.
- *Messaging service agent*: the access point to the messaging service for suppliers and consumers. It controls and co-ordinates all channels within the messaging service.
- *Service management system agency*: a particular agency which hosts the MSA. Supplier agents and consumer agents can also be hosted.

These different computational objects are shown in Figure 9.17 and described below.

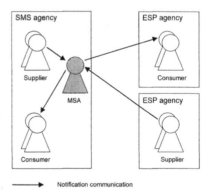

Figure 9.17 Overview of the messaging service architecture

**SMS agency**
The SMS agency is a specific agency within the MARINE environment that provides access points to management capabilities for administrators. Within the NFD, only one agency has the privilege to act as the SMS agency. The MSA resides on a specific place within this agency, the so-called management place.

---

[4] In the MARINE context, a single notification exchange domain is sufficient to fulfil the requirements of the services defined.

**ESP agency**

The ESP agency is located within the MARINE environment where the service provisioning takes place. There can be multiple instances of an ESP agency.

**MS agent**

The MSA provides the functionality of fulfilling the requirements identified within the previously defined use case model. The MSA is instantiated during the set-up of the SMS. The decision, where the MSA is created, must be determined off-line by the administrator of the NFD. This person has also to guarantee that only one instance of an MSA exists within the NFD.

Once created, the identity and location information of the MSA is published. That means these two pieces of information are provided to a wellknown and accessible location, such as a Naming Service. Thereafter, the MSA is ready to provide the notification forwarding mechanism to suppliers and consumers. During the service provisioning, the MSA must handle the creation, deletion, suspension, and resumption of channels, as well as the management of supplier and consumer related information such as identities, access points, and policies. The supplier and consumer information is collected during the registration process (see *Open Channel* and *Attach to Channel* use case).

For the actual distribution of notifications, the MSA must evaluate all the policies related to a particular channel in order find out the consumers of a particular notification. In accordance with the number of consumers, the MSA has to duplicate the notifications and multicast them.

Besides the provision of an access point to the notification forwarding mechanism, the MSA provides a management access point for an administrator to monitor and control the MSA itself. This access point qualifies the administrator to view all opened channels, their states, as well as supplier and consumer related information. Additionally, some management actions such as an administrative order to remove, suspend, and resume a set of channels are handled by this agent.

**Supplier agent**

The supplier is an agent, acting as a supplier as described within the use case model. This agent emits notifications to the MSA instance. For this purpose, the supplier provides its identity and optionally the NFP to the MSA for registration purposes and to create a new channel. Suppliers can reside on any agency within the NFD for emitting notifications to the MS agent. The access point to the MSA can be retrieved from the naming service.

The creation of a channel is not dependant on the existence of consumers. The supplier will not be informed by the messaging service whether consumers are attached to or detached from the channel. If a supplier is constituted with the knowledge that only some consumers will receive notifications, it can customize its channels with the notification forwarding policy.

A supplier will be aware of the fact that a channel can be closed by an administrator. The supplier does not realize this directly; it will only become aware that its access point is not valid anymore. Therefore, a supplier must be endowed with a reaction behavior to the closing of a channel through an administrator.

**Consumer agent**

The consumer represents an agent that consumes the emitted notifications from suppliers just by attaching to the supplier's channel. To access the channel, the consumer must retrieve the access point to the MSA via the Naming Service. Multiple consumers can be attached to one or multiple suppliers. The location of the suppliers does not matter at all, they can reside on the same agency as well as on a remote one, since the actual forwarding mechanism is hidden inside the MSA and a consumer is not directly connected to the supplier at all.

An attachment to a channel will only be successful for a consumer if the targeted channel is already opened by the supplier. As stated before, a channel is always associated with a single supplier. Nevertheless, the knowledge from which suppliers a consumer wants to receive notifications must be acquired prior to the attachment to the channel. This knowledge can be a constituent part of a consumer, injected during the agent design phase.

As the supplier, a consumer can also customize the channel by subscribing the types of interesting notifications. In case a supplier closes a channel even though consumers are still attached to it, each consumer receives a predefined notification from the MSA. The consumer reacts to this notification by doing some clean ups. This also applies if the channel is closed by the administrator.

### Channel

The channel describes a logical link between a supplier and a set of consumers. It defines the path from a supplier to the MSA, and from there on to all consumers attached to that channel. Hence, it expresses the possible flows of notifications from a supplier to consumers.

Both the notification forwarding and receiving policy are associated with a channel. These policies provide constraints that regulate the notification forwarding mechanism.

## 9.6 Designing and Implementing IN Network Elements within the DOT/MAT Environment

Network elements within the DIBN are implemented on the basis of Intelligent Broadband Network systems. In particular, two systems are considered as a basis for the DIBN:

- B-SSPs, where service triggers are detected, and support to requests made by Service Logic Programs in terms of connectivity is provided.
- B-SCPs, providing centralized service handling to a multiplicity of B-SSPs.

Enhancements to these systems are made to obtain the DIBN network elements, the key objective being the possibility to move services from centralized elements to distributed ones, depending on service needs.

Two kinds of enhancements are considered. The first is a technology upgrade, and is related to the integration of DOT and MAT platforms within the core system architecture. The second enhancement is related to the development of wrappers based on DOT and MAT, enabling legacy system functionality (as an example, B-SSF within a B-SSP) to access the distributed environment.

The DIBN systems derived from the B-SSP and the B-SCP are B-SS&CP and SEN, respectively (see Section 9.3) (Venieris, 1998). These systems will be described in detail in the following, considered their physical architecture, technological components and functionality.

## 9.6.1    Broadband Service Switching and Control Point

The B-SS&CP consists of an IN B-SSP enhanced with a B-SCF/B-SDF functionality, allowing it to locally run IN services. This system is connected to the SMS, from which it downloads service logic prototypes, and to a Service Execution Node, to which requests to infrequently accessed services are directed (Faglia, 1998).

The B-SS&CP is physically composed by three separate subsystems:

- ATM switching matrix
- Signaling Server
- Services.

The first two subsystems are inherited by a B-SSP prototype, while the third one is an innovation needed for the DIBN; and is used to run services directly within the B-SS&CP.

The *ATM switching matrix* subsystem is an ATM cross-connect offering a control interface allowing the creation/deletion of cross-connection between different ports.

The *Signaling Server* subsystem runs signaling and IN protocol stacks and sets up/tears down connections on the *ATM switching matrix* according to calls made by users (or committed by IN network elements). Provided that a suitable control interface is offered, the Signaling Server can use a variety of switching matrices provided by different vendors. This subsystem consists of a VME board controlled by a microprocessor, with dynamic RAM, Ethernet and SCSI controllers, and equipped with an ATM module. In the experimental configuration, a commercial board equipped with a PowerPC 604/100MHz processor and 32 Mbytes of dynamic RAM is used. The VxWorks® real-time kernel provided by WindRiver Systems controls the board.

The Signaling Server is an experimental prototype running signaling/IN protocols developed by Italtel Central R&D and written in C/C++ (Venieris, 1998). Standard signaling protocols as specified by ITU-T and ATM Forum are used on UNI and NNI interfaces. A proprietary B-INAP protocol running on top of a standard SS7 on ATM protocol stack is also used (Prezerakos, 1998). As an alternative to B-INAP, the Signaling Server can communicate with B-SCF functionality by means of DPE-based communication protocols. Due to a limitation of the selected commercial DPE platform, only a proprietary DPE protocol can be used. This constraint derives from the fact that Signaling Server applications are written in C/C++, and no C/C++ CORBA-compliant ORB was available at implementation time for the VxWorks® environment.

The *Services* subsystem is physically identical to the Signaling Server but hosts middleware (CORBA and MAP) platforms and services downloaded from the network. This third subsystem is completely Java-based. The VxWorks® real-time kernel is again used, and on top of that a *Java Virtual Machine* (JVM) provides the runtime environment for CORBA and MAP platforms and related services. The adoption of a Java runtime environment allowed the use of a Java ORB. In particular, to ease CORBA-based communication between Signaling Server and Services subsystems, a product by the same vendor providing the ORB for the Signaling Server and completely compatible with CORBA 2.0 specifications was chosen.

To provide standard CORBA communication, based on the IIOP protocol, between the B-SS&CP and other DIBN systems, and considering the limitations of the Signaling Server subsystem, a protocol bridge has been included in the design of the Services subsystem.

This bridge maps requests coming from Signaling Server and based on the proprietary DPE protocol on standard IIOP ones, and vice versa. Therefore, all IN service requests made by the Signaling Server traverse the Services subsystem. Requests being directed to locally available services are handled within Services, while external ones are routed to the proper external server. Figure 9.18 shows the physical configuration of the B-SS&CP.

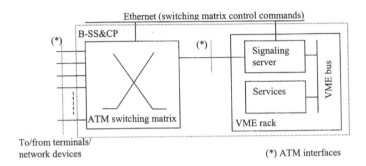

**Figure 9.18**  B-SS&CP physical architecture

Users and network devices are connected to ports of the ATM Switching Matrix. Signaling information used on these ports is cross-connected to a dedicated port, which is in communication with the Signaling Server (Venieris, 1998).

Signaling Server and Services subsystems communicate by means of the VME bus interface. Service agents coming from the network enter the B-SS&CP via one of the ATM interfaces and are relayed to the Services subsystem by means of routing functionality provided by the Signaling Server. Communication between the Services subsystem and other entities in the network is again performed through routing made by the Signaling Server.

The number of boards used for Signaling Server and Services subsystems can vary according to performance requirements. Software functionality supported by these boards can be allocated to different processors in order to increase throughput.

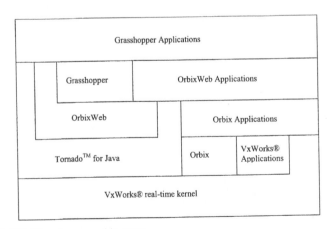

**Figure 9.19**  B-SS&CP software architecture

From the software viewpoint, signaling/IN and Service subsystems of the B-SS&CP are seen as a unified entity. Their software architecture is layered as reported by Figure 9.19.

As Figure 9.19 shows, the following technological choices were made:

- ORB for Signaling Server subsystem: Orbix by IONA Technologies
- Java Virtual Machine for Signaling Server subsystem: TornadoTM for Java
- ORB for Services subsystem: OrbixWeb by IONA Technologies
- MAP platform: Grasshopper by IKV++.

**VxWorks® real-time kernel**

VxWorks® provided by Wind River Systems is the real-time operating system on which the software architecture is based. Version 5.3.1 is used in the prototype. This kernel provides the execution framework for:

- The TornadoTM for Java JVM
- Orbix
- VxWorks® applications, which include signaling stacks and IN applications.

**Tornado$^{TM}$ for Java**

The Java Virtual Machine provided by Wind River Systems allows Java applications to be executed under the VxWorks® OS. A single instance of the JVM is supported, and Java applications are started as threads. The version of this JVM used for the B-SS&CP prototype supports JDK 1.1.4. The Tornado$^{TM}$ for Java JVM supports:

- OrbixWeb
- Grasshopper
- OrbixWeb applications
- Grasshopper applications.

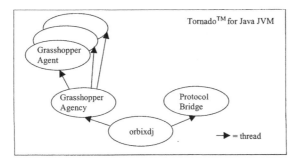

**Figure 9.20**  Usage of the Tornado$^{TM}$ for Java single instance JVM

Due to this limitation of the JVM, and considering that the OrbixWeb activation daemon has to be used to support proprietary ORB communication between Signaling Server and Services, the following approach was selected:

- The JVM instance is taken by the OrbixWeb activation daemon

- A Grasshopper agency and the protocol bridge are started as 'in process' servers[5]
- All Grasshopper agents are started as Java threads.[6]

### Orbix

The real-time version of Orbix provided by IONA technologies is used here to provide CORBA encapsulation of SSF and MCUSF functionality offered by the B-SS&CP. The version of this ORB used in the prototype is Orbix 1.3.5, which has some limitations with respect to the capabilities offered by the most advanced versions of Orbix, including the following:

- IIOP is not supported
- The Naming Service is not supported
- Only persistent servers are available.

Orbix supports wrapping of SSF and MCUSF functionality into CORBA.

### OrbixWeb

The Java version of IONA's ORB is used to support SCF functionality within the B-SS&CP. Version 3.1 is used in the prototype.

The OrbixWeb activation daemon is started within the single instance of the Tornado for Java JVM, and CORBA servers are started 'in-process' as Java threads. OrbixWeb offers support to:

- OrbixWeb applications
- Grasshopper
- Grasshopper applications.

### Grasshopper

Together with OrbixWeb, Grasshopper provided by IKV++ is the basis for the SCF functionality offered by the B-SS&CP. Version 1.2 is used in the B-SS&CP prototype.

Grasshopper supports an agency within the B-SS&CP dedicated to retrieving services from the network and running them on the B-SS&CP itself.

### VxWorks® applications

The following functionality is considered as part of the set of VxWorks® applications in the B-SS&CP context:

- CCF
- SSF
- MCUSF
- Switch handler
- Transport.

This functionality is structured according to Figure 9.21.

---

[5] CORBA servers can be started 'in process' as a feature of OrbixWeb. This implies running them as Java threads started by the OrbixWeb activation server (orbixdj).

[6] This is the standard way Grasshopper agents are created within a single agency.

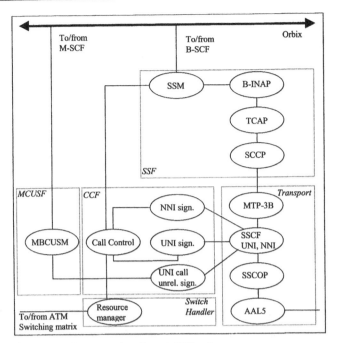

**Figure 9.21** VxWorks® applications: signaling and IN software

The VxWorks® real-time kernel obviously also supports middleware platforms like Orbix, OrbixWeb and Grasshopper and related applications. For the sake of simplicity, only applications directly accessing OS functionality without the intervention of a middleware layer are named 'VxWorks® applications' in this context.

VxWorks® applications communicate by means of *Inter Process Communication* (IPC) facilities provided by the VxWorks® OS. IPC functionality is not directly accessed by the applications; an intermediate custom library is used instead, to hide OS-specific functionality to the applications themselves. This considerably facilitates porting of this software to other OSs if and when necessary.

**Orbix applications**
The B-SS&CP hosts the following Orbix applications:

- `AccessPointToServiceKeyHandler`, `AccessPointToServiceLogic` objects belonging to the SSM, which are responsible for handling the CORBA-based interaction with service logics via the SSF Access Manager (part of the OrbixWeb applications). These CORBA objects are related to the main SSM structure, as shown at a high level in Figure 9.22
- `AccessPointToMobilityKeyHandler`, `AccessPointToMobilityServ-Logic` objects belonging to the MBCUSM, responsible for handling the CORBA-based interaction with Mobility Management procedures via the MBCUSM Access Manager (part of the OrbixWeb applications). Relationships between these CORBA objects and the main MBCUSM body are given in Figure 9.22.

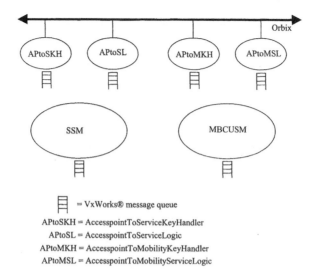

**Figure 9.22**  High level view of Orbix applications

The Orbix applications are related to VxWorks® ones, since they act as the 'liaison' between native signaling/IN software of the B-SS&CP and the new DOT/MAT based Services subsystem of the switch. The relationship takes between the two kinds of applications place via IPC communication provided by VxWorks®.

**OrbixWeb applications**
The only 'pure' OrbixWeb application hosted by the B-SS&CP is the protocol bridge. The bridge actually consists of a set of CORBA servers without relationships with agent technology.

**Grasshopper applications**
The B-SS&CP hosts a number of Grasshopper applications related to the support of B-SCF and M-SCF functionality. Both Grasshopper stationary and mobile agents are present. Stationary agents are obviously always present in the system, while mobile agents depend upon services locally supported by the B-SS&CP in a given time frame. Five Grasshopper stationary agents are available:

- Bootstrapping Agent: this agent performs the start-up of the remaining stationary agents and takes care of interconnecting them.
- SSF Access Manager, which provides access to B-SCF agents.
- Service Logic Manager-Stationary Part, which controls the instantiation of service logics (both local and remote).
- MBCUSM Access Manager, which provides access to M-SCF agents.
- Mobility key handler-stationary part, which controls the instantiation of mobility management agents (both local and remote).

In addition to stationary agents, the mobile parts of the Service Logic Manager and Mobility Key Handler are always made available locally to the B-SS&CP by the SMS. These agents maintain information about known services/mobility management procedures.

Other B-SCF and M-SCF Grasshopper applications implemented as mobile agents can be made available locally to the B-SS&CP, depending on service needs.

## 9.6.2    Service Execution Node

From the viewpoint of IN physical elements, the SEN can be regarded as the integration of B-SCP and B-IP functionality into a single *Service Node* (Prezerakos, 1998). The SEN receives service logic prototypes from the SMS, and is connected to a multiplicity of B-SS&CPs in order to:

- Run IN service logic programs in order to comply with service requests coming from users via the B-SS&CP, thereby instructing the switches in order to establish/tear down bearer connections according to service specific characteristics.
- Migrate service logic prototypes to the B-SS&CP, on the basis of explicit requests made by the switches themselves depending on a programmable service access threshold.

The hardware basis used for the SEN is a Sun workstation running Solaris 2.5.1 (or higher) and hosting an ATM adapter. Within the experimental configuration, an UltraSparc workstation equipped with 128 Mbyte dynamic RAM, 3 Gbyte HDD and a Fore SBA-200E/OC3ST ATM adapter is used.

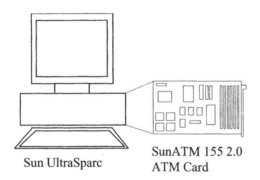

Sun UltraSparc              SunATM 155 2.0
                            ATM Card

**Figure 9.23**   SEN physical architecture

From the software viewpoint, the SEN makes use of a commercial basis on top of which advanced IN Service Logic Programs for SCF and SRF functional entities run.

The same middleware platforms used in the B-SS&CP are adopted for the SEN as well. Concerning signaling and SS7 protocol stacks, they are provided respectively in order to:

- Allow the SEN to receive calls coming from users. The SEN uses bearers established with these calls in order to transfer service menus to the users themselves, and to receive service related information from them. In this case, the SEN behaves as a B-IP.
- Receive IN service requests coming from B-SS&CPs (and/or traditional B-SSPs) via the SS7 stack.

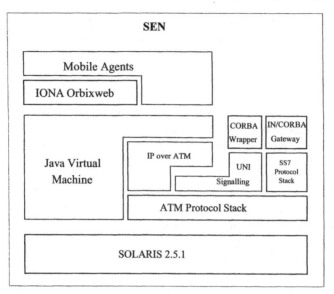

**Figure 9.24**  SEN software architecture

**Signaling**

The Signaling stack can be divided into the SAAL and the *Access Signaling* (AS). The SAAL consists of the Service *Specific Connection Oriented Protocol* (SSCOP) and the SSCF.

The SSCOP provides its user (the AS) with an assured message transfer over the underlying non-reliable layer which is the AAL5. The SSCF resides above the SSCOP and simplifies the interface between the SSCOP and the upper layer of the UNI stack. Above the SAAL the UNI 3.1 (ITU-T/Q.2931, 1995) enhanced with Mobility Management procedures (ITU-T/Q.2932, 1996) signaling stack is implemented.

The *Access Signaling* CORBA server provides the CCAF object with the VPI, VCI values that correspond to a specific `SessionId`. The wrapper and the AS stack communicate with each other via a Socket interface. The wrapper is a CORBA server providing a generic method, taking as parameters the name of the operation and its arguments. These parameters are afterwards passed to the AS stack where they are processed, and the result is passed to the wrapper and to the CORBA client. The server is implemented in C++ while the CCAF client is in Java.

The AAL5 is also CORBArized. Two CORBA wrappers have been developed, the AAL5Send and the AAL5Receive. The former acts as a server to the DCA, while the latter acts as a client. Both of them are implemented in C++.

The AAL5Send wrapper first initiates the outgoing ATM *Virtual Channels* (VCs) identified by the VPI/VCI pair specifying the exact values of QoS and other connection characteristics, using the `atmconnect` method. Afterwards, the AAL5 packets can be sent to these VCs with the `atmsendto`. It also provides *Segmentation And Reassembly* (SAR) functionality conveying packets of arbitrary length to AAL5 ones.

The AAL5Receive wrapper acts as a client to the DCA. First, it initiates the incoming ATM VCs identified by the VPI/VCI values specifying the QoS reserved for them, using the `atmbind` method. All traffic received from these VCs is available to the wrapper, via the `atmrecvfrom` method. This traffic is then conveyed to the DCA.

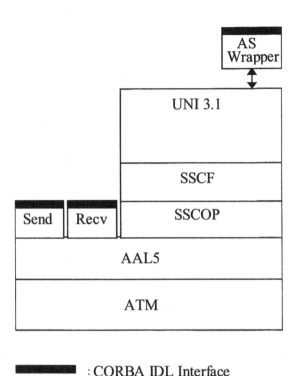

: CORBA IDL Interface

**Figure 9.25**   Signaling stack and CORBA wrappers

**SS7 protocol stack**

The SS7 protocol performs layers starting from SSCOP up to B-INAP, both included. The ATM adapter on which the stack is based performs the three lower layers of the protocol stack (i.e. SDH/STM1, ATM and AAL5).

This version of the stack uses the FORE SBA200E ATM adapter, and runs on top of the A_ForeThought_4.1.0 (1.71) driver. The two processes (`MapperDown_fore`, `MapperUp_fore`) perform the adaptation to the ATM card and related driver. This is the only part of the protocol stack depending on the underlying hardware.

On top of the SS7 stack, the IN/CORBA gateway is run. The gateway allows MAT- and DOT-based service logic programs running in the SEN to receive service requests through SS7. By means of the gateway, the nature of service requests is hidden to the SLPs, which behave as services were requested by means of CORBA. The gateway implementation architecture is reported in Figure 9.27.

All gateway objects have been written in C++. Orbix 2.3 provided by IONA Technologies was used.

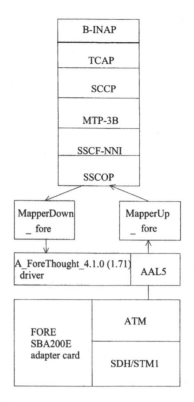

**Figure 9.26**  SS7 protocol stack

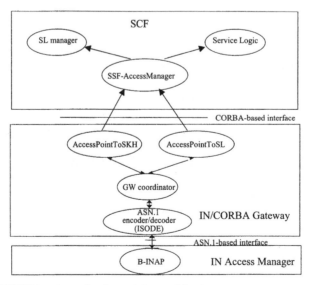

**Figure 9.27**  IN/CORBA gateway implementation architecture

Figure 9.27 also shows how the gateway is related to the other SEN modules. Two external interfaces are implemented:

- An Abstract Syntax Notation 1 (ASN.1)-based interface, which makes use of IPC functionality in order to allow the gateway to communicate with the B-INAP protocol stack.
- A CORBA-based one, which is used to communicate with service logics hosted by the SEN via the SSF access manager.

The second interface has been designed to be identical to that connecting the (`AccessPointToServiceLogic`, `AccessPointToServiceKeyHandler`) objects in the B-SS&CP to the SSF Access Manager. This allows service logics (and the SSF access manager) to be independent from their execution environment (i.e. the same code runs in the SEN and in the B-SS&CP).

# References

Breugst, M. and Choy, S. (1999) Management of Mobile Agent Based Services, Lecture Notes of Computer Sciences 1597, Intelligence in Services and Networks: Paving the Way for an Open Service Market, Han Zuidweg et al. (Eds.), pp. 143–154, *Proceedings of 6th International Conference on Intelligence in Services and Networks (IS&N)*, ISBN: 3-540-65895-5, Springer-Verlag 1999, Barcelona, Spain.

Faglia, L., Marino, G. and Zizza, F. (1998) The interworking of IN and B-ISDN and the introduction of mobile agents technology. *Proceedings of Interworking '98*, Ottawa.

Herzog, U. and Magedanz T. (1998) Intelligent Networks and TINA – Migration and Interworking Issues, The Intelligent Network: Current Technologies, Applications, and Operations, pp. 109–122, ISBN: 0-933217-43-9, International Engineering Consortium (IEC), Chicago, IL.

ITU-T, Recommendation Q.2931 (1995) Broadband Integrated Services Digital Network (B-ISDN) – Digital Subscriber Signaling System No. 2 (DSS 2) – User-Network Interface (UNI) – Layer 3 specification for basic call/connection control.

ITU-T, Recommendation Q.2932.1 (1996) Digital Subscriber Signaling System No. 2 – Generic functional protocol: Core functions.

ITU-T, Recommendation X.701 (1992) ISO/IEC 10040: 1992. Information Technology – Open Systems Interconnection – Systems Management Overview, International Standard.

ITU-T, Recommendation X.720 (1993) ISO/IEC 10165-1: 1993. Information Technology – Open Systems Interconnection – Structure of Management Information: Management Information Model, International Standard, 1993.

ITU-T, Recommendation X.722 (1992) ISO/IEC 10165-2: 1992. Information Technology – Open Systems Interconnection – Structure of Management Information: Guidelines for the Definition of Managed Objects, International Standard, 1992.

Magedanz, T., Rothermel, K. and Krause S. (1996) Intelligent Agents: An Emerging Technology for Next Generation Telecommunications? *Proceedings of IEEE INFOCOM '96*, pp. 464–472, IEEE Catalogue No. 96CB35887, ISBN: 0-8186-7292-7, IEEE Press.

OMG, CORBA Facilities Architecture Specification (1998) http://www.omg.org/corba/cf2.html.

Prezerakos, G., Pyrovolakis, O. and Venieris, I.S. (1998) Service Node Architectures incorporating Mobile Agent Technology in CORBA and Broadband IN. *Proceedings of IFIP/IEEE MMNS'98*

Versailles-France.

Prezerakos, G., Salsano, S., van der Vekens, A. and Zizza, F. (1998) INSIGNIA: a pan-European trial for the Intelligent Broadband Network architecture. *IEEE Communications Magazine*, **36**(6).

TINA-C (1997) Service Architecture, Annex, TINA-C Baseline Document.

Venieris, I.S. and Hussmann, H. eds. (1998) *Intelligent Broadband Networks*. John Wiley & Sons Ltd.

# 10

# Service Specification in the Distributed Intelligent Network

## 10.1 Service Description Methodology: UML

The development of the *Object Oriented* (OO) programming paradigm started in the early 1980s, and in the meantime, object orientation has become a wellaccepted technique in a large area of software applications. Different methodologies have been provided from the beginning on, realizing guidelines for the design and development of software systems, such as *Object Oriented Analysis* (Coad, 1991a) and *Object Oriented Design* (Coad, 1991b) from Coad/Yourdon, *Object Modeling Technique* (OMT) (Rumbaugh, 1991) from Rumbaugh, *Object Oriented Analysis and Design with Applications* (OOADA) (Booch, 1994) from Booch, *Object Oriented Software Engineering* (OOSE) (Jacobson, 1992) from Jacobson, and many others. These methodologies differ in their approach, level of detail, basic concepts, and terminology. Due to this lack of agreement, problems arose for users as well as for vendors of available OO tools. The considerable costs of using and supporting many different modeling languages motivated numerous companies to push towards the development of a single, unified language. The product of this development effort is the *Unified Modeling Language* (UML).

The purpose of UML is to support the specification, implementation, and also documentation of complex software systems.

### 10.1.1 Overview

The Unified Modeling Language provides a notation for object oriented software development, combining and unifying the notations of OMT, OOADA, and OOSE. UML can be used to:

- Specify *use cases* in order to define the basic functionality and boundaries of a software system.
- Describe the static system structure by means of *class diagrams*.

*Object Oriented Software Technologies in Telecommunications*, Edited by I. Venieris, F. Zizza and T. Magedanz
© 2000 John Wiley & Sons Ltd

- Describe the behavior of objects in terms of *state diagrams.*
- Explain the dynamic behavior of the system by means of *interaction diagrams.*
- Specify the architecture and implementation with *component* and *deployment diagrams.*
- Enhance the provided notation and semantics by means of *stereotypes, notes, tagged values, patterns,* and *constraints.*

After several modifications, the Object Management Group (OMG) distributed the standardized version 1.1 of the UML specifications (OMG/UML1.1, 1997). The actual UML specification (OMG/UML1.3, 1999) is available for free via the World Wide Web. In particular, it describes:

- UML Semantics, comprising a formal definition of UML.
- UML Notation Guide, describing the graphical and textual UML notation.
- UML Extensions, explaining how to realize user-defined UML extensions by means of stereotypes, tagged values, and constraints.

The following sections briefly address the types of UML diagrams that can be used to describe a system in terms of different views.

## *10.1.2 Use case diagrams*

Use case diagrams describe the problem domain of a software system, covering the system's basic functionality as well as the communication between actors and the system. Actors represent external systems or human users. Since they are always situated outside the system, use case diagrams help to define the boundaries of a system.

## *10.1.3 Class diagrams*

A class represents a set of relationships and semantics, and class diagrams can be used to describe the static structure of a system. Each class may provide a set of attributes and operations and use a set of interfaces, specifying collections of operations it provides to the environment.

    Classes and their relationships describe the realization of use cases in a static manner. Relations between classes symbolize generalization/specialization, association, aggregation, and dependencies.

## *10.1.4 Behavior diagrams*

A system's behavior can be expressed in terms of three different diagrams:

- *Statechart diagrams* that describe the behavior by means of state machines, including simple states, transitions, and nested composite states.
- *Activity diagrams* that define the behavior in terms of activity graphs, which are a special kind of state machines. These state machines consist of activity states, representing the execution of an internal computation. In contrast to usual states, activity

states should imply transitions due to the completion of the internal computation, rather than due to explicit, external events.

- *Interaction diagrams* describe interactions between objects, where each object is an instance of a class. In this way, interaction diagrams provide dynamic aspects to the static class diagrams. UML defines two kinds of interaction diagrams: sequence diagrams, showing a set of messages arranged in a time sequence, and collaboration diagrams, arranging the objects and links in a geometric manner.

## 10.1.5  Implementation diagrams

Implementation diagrams describe a system in terms of components which are physical units of implementation, realizing replaceable parts of the system, and offering well defined interfaces. Each component covers a set of classes from the system design. UML supports two different kinds of implementation diagrams:

- *Component diagrams* which consist of a set of components and depict their dependencies. A dependency is symbolized by a connection from one component to the interface of another component.
- *Deployment diagrams* which cover a set of component instances and their dependencies. Note that, in contrast to the component diagram, the deployment diagram takes into account the distribution of the instances across the environment. For this purpose, node instances are introduced, representing specific locations within the environment.

## 10.1.6  UML usage in MARINE

Within the framework of the MARINE project, a subset of UML has been adopted as common notation for the design and specification of the entire software environment. The purpose was to achieve a unified system design and deployment in order to ease the integration of components realized by different project partners.

The following subset of UML has been used in MARINE:

- *Use case diagrams*, which provide an abstract overview of the functionality of the MARINE services and the surrounding infrastructure. They have been defined according to the results of the requirement analysis, performed at the beginning of the project.
- *Class diagrams*, used to describe the static composition of the MARINE services and of the infrastructure by focusing on interfaces between major components, developed by different providers.
- *Collaboration* and/or *sequence diagrams*, used to capture the dynamic behavior of the MARINE software. The purpose was to ensure that all fundamental interaction scenarios could be performed with the developed classes.

The following sections outline a subset of the MARINE service specification, describing the software in terms of use case diagrams and collaboration diagrams.

Depending on the objects' characterization and distribution, the communication mechanism between objects within collaboration diagrams is realized in terms of signaling interactions, internal message exchange (inter-process communication, local Java method invocations) or remote CORBA requests.

On the other hand, it has to be noted that concrete implementation components and their distribution onto the different network nodes of the MARINE infrastructure have not been described in terms of UML implementation diagrams. Instead, custom notations were used that differentiated between component types that are of particular importance in MARINE, such as mobile/stationary agents, agencies, CORBA servers, etc.

## 10.2   IMR Service

*Interactive Multimedia Retrieval* (IMR) is a service that provides transfer of digitally compressed and encoded video and audio data across a telecommunication network. The information is retrieved on the user's demand and on an individual basis. In fact, the user is given the possibility to choose the exact time at which the information stream is to be transmitted and to exert overall control during the transmission itself.

The information is stored, categorized and classified by one or more *Service Providers* (SPs) which use *Multimedia Servers* or *Video Servers* (VS) to offer the IMR service to the users. The destination device or *Customer Premises Equipment* (CPE) is usually a *Set Top Box* (STB). The STB is connected to a TV set and has a broadband access to the network. An ATM PC or a *Network Computer* (NC) can also be used as CPE.

Service selection is achieved by means of an on-screen menu. The user's commands are relayed across a network to the IMR Video Server via a Selection Path channel, while the video and audio data is transmitted via a Multimedia channel.

The interactive part of the selection phase is called 'navigation'; two levels of navigation are provided, the first one handling the Service Provider and Sub-service selection and the second handling the Content selection. The entity receiving from the user the information regarding the selected Service Provider and Sub-service is called *Broker*.

The first level of navigation includes the *authentication* of the CPE, the *authorization* of the user, and the *selection* of a SP and a sub-service. Afterwards, the second level of navigation takes place, where the user selects the content from the available IMR Video Server that belongs to the selected SP. The Video Server acts as a content repository and, on request, delivers the appropriate content through a broadband channel (downstream channel) with bandwidth modifiable upon demand, down to the STB.

The user is given interactive control of the playback, identical to VCR control commands, i.e. play, pause/resume, fast forward/fast backward.

Some features of a commercial IMR service are omitted; in particular, charging and billing aspects are not covered.

### 10.2.1   IMR use case model

As already introduced in Section 10.1.2, a use case diagram shows the relationships between the actors and the use cases of the system. It can also show the relationship between use cases of the system itself.

With respect to the MARINE service specification, the use case can be initiated either by the actor or by the system.

In particular, concerning the case study applied to the Interactive Multimedia Retrieval Service (Figure 10.1), there exists one actor – the originating user/STB (a service provider being the terminating user/IMR Server can be regarded as belonging to the system itself) and two use cases: the first one initiated by the actor (*IMR Call Setup – User Initiated*), the second one initiated by the system/Network *(IMR Call Setup – Network Initiated)*.

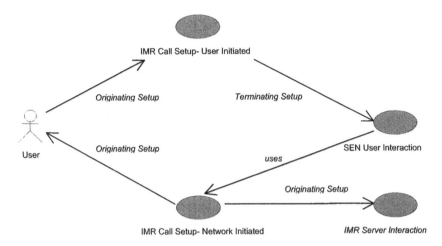

**Figure 10.1** IMR case diagram

In the first use case, which represents the beginning of an IMR transaction, a service session is always started by the user/STB by generating a Signaling message (Originating Setup) towards the Broadband-Service Switching and Control Point and containing the IN number associated with the IMR service.

The B-SS&CP recognizes the IN trigger for the new IMR service session and asks the IMR Service Logic, which is distributed on the SEN, for the necessary routing information (the STB must be connected to the SEN acting as a Broker for the first-level navigation).

After the reception of the routing information (Called Party number), the B-SS&CP sends a (Terminating) Setup message towards the SEN for establishment of the connection associated with the first-level navigation.

During this phase (SEN User Interaction), Authentication/Authorization procedures and SP selection are carried out through actions performed by appropriate SL modules.

After the SP has been selected, the call towards the SEN is released for the second level of navigation. In fact within the second use case which continues the IMR transaction, two connections (for the Selection Path channel and Audio/Video stream) between the user and the IMR Server are established through Network-Initiated calls by means of the IMR SL on the Service Execution Node. Finally, the calls towards the IMR Server are released by the user and the Set Top Box is re-connected to the SEN for a new (eventual) SP selection or release of the service session.

## 10.2.2 IMR objects characterization

According to the different levels of abstraction provided by the UML methodology, the dynamic behavior of the system realizing the Interactive Multimedia Retrieval service can be well characterized by object interactions depicted in related collaboration diagrams.

For this purpose, it is necessary to introduce the main objects of the system that are involved within the above-mentioned interactions.

**Table 10.1**  Main objects implementing the interactive multimedia service

| Object | Operation |
| --- | --- |
| AccessPointToSKH | The `AccessPointToServiceKeyHandler` object is responsible for starting a service session from the B-SS&CP side by invoking a request (`ServiceKeyRequest` with an appropriate `ServiceKeyNumber`) to the `SSMAccessManager` object, which in turn, will submit the request to the `ServiceLogicManager` object. |
| AccessPointToSL | The `AccessPointToServiceLogic` object handles the dialogue (B-SS&CP side) with the Service objects (IMR service sessions) through mediation of the `SSFAccessManager` object. |
| SSFAccessManager | The SSF Access Manager object can be considered as an 'intermediary' between SSF and SCF Functional Entities. Each method invocation issued by the `AccessPointToSK` and `AccessPointToSL` objects is processed by the `SSMAccessManager` object. |
| ServiceLogicManager | The SLM is responsible for resolving the service from the incoming `ServiceKeyNumber` and triggering the appropriate object (`IMRService`) accomplishing IMR SL tasks. |
| IMRService | The `IMRService` object represents the open IMR service session between end users. The `IMRService` object is created (and instantiated) by the `ServiceLogicManager`. |
| InnerSSM | 'Internal SSM objects' that encompass the different modules located on the B-SS&CP equipment, which functionality is to handle the Switching State Model. |
| DispatcherManager | The interaction with the `DispatcherManager` object has to be established in order to determine the communication mode (INAP or CORBA). |
| AccessPointToLegBCSM | The `AccessPointToLegBCSM` object allows the access to the Call Control process residing on the B-SS&CP. |
| CallControl | This object (residing on the B-SS&CP) represents the point of interface between B-ISDN signaling and IN. It realizes the Basic Call State Model (BCSM), and is responsible for establishing and maintaining (through signaling) communication paths between users. |
| OriginatingCallControlAgentFunction | This object is responsible for originating the service call. Within IMR service scenarios the Originating CCAF is located on the Set Top Box. |

| Object | Operation |
|---|---|
| `TerminatingCallContr olAgentFunction` | This object is responsible for controlling an incoming call. Subsequent diagrams can be located on the Service Execution Node (Terminating SEN), Set Top Box (Terminating STB) and IMR Server (Terminating IMR Server). |
| `TerminalProfile` | The use of the Terminal Profile object for IMR is basically bound to the storage of the Terminals (Set Top Box, IMR Server, SEN acting as a Broker) E164 numbers. |

## 10.2.3 IMR service description

Starting from the previously mentioned objects and according to the use case defined in Section 10.2.1, in the following the IMR Service is presented in terms of collaboration diagrams and actions performed on the objects for the purpose of describing just the first use case: the establishment of the connection between user/STB and SEN/Broker.

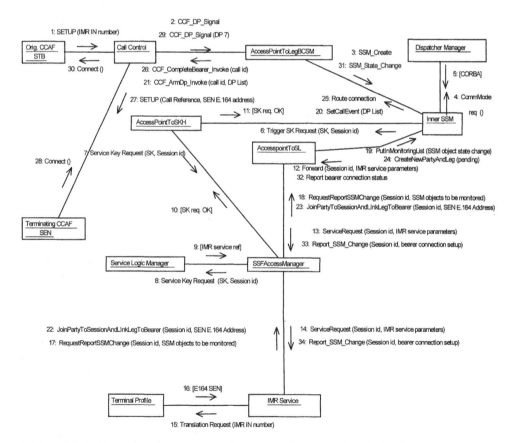

**Figure 10.2** IMR collaboration diagram – the user/STB is connected to the SEN

**Table 10.2** Actions performed by the objects during STB/SEN connection establishment

| Message | Action |
| --- | --- |
| 1. `SETUP` | Signaling message sent to the B-SS&CP by the Call Control functionality (Orig. CCAF) residing on the STB/user-terminal (Calling Party). |
| 2. `CCF_DP_Signal`,<br>3. `SSM_Create` | B-SS&CP internal communication mechanism between CCF and SSF functionality indicating (2) to the last one that an IN Trigger for an IMR service request detected by the Call Control object has to be handled by the Inner SSM object. For this purpose this object creates (3) and manages an Switching State Model (SSM) data structure to handle (through opportune SSM state changes) all service session related events. |
| 4. `CommModeReq`,<br>5. `[CORBA]` | B-SS&CP internal communication between the SSF functionality and the Dispatcher Manager object in order to identify through an opportune request (4) the communication mode to be established with the SEN equipment; via CORBA (5) or via INAP protocol. Within this context, CORBA communication is considered. |
| 6. `TriggerSKRequest` | B-SS&CP internal communication between the Inner SSM object and B-SS&CP objects implementing CORBA functionality (`AccessPointToSKH`, `AccessPointToSL`). This message activates the `AccessPointToSKH` object for a subsequent method invocation corresponding to the `ServiceKeyRequest`. |
| 7, 8. `ServiceKeyRequest` | CORBA method invocation. By invoking this method on the `SSFAccessManager` server object the client (`AccessPointToSKH`) initiates an IMR service session including the appropriate `ServiceKeyNumber` and session identifier (`SessionId`) as method parameters. The `SSMAccessManager` object in turn, acting as a client relays this method invocation towards the `ServiceLogicManager` object. When this method returns, the `SSFAccessManager` maps ServiceLogic instance to `SessionId`. The `ServiceLogicManager` generates a new IMR Service instance by cloning a mobile agent that serves as a prototype implementation for that service. |
| 9. `[IMRServiceRef]`,<br>10, 11. `[SKReqOK]`,<br>12. `Forward` | B-SS&CP internal communication. After creating the Service instance the `ServiceLogicManager` returns an IMR service object reference to the `SSFAccess-Manager`. On successful invocations |

| Message | Action |
|---|---|
| | [SKReqOK] the `InnerSSM` object Forwards `SessionId` and necessary IMR service parameters to the `AccessPointToSL` object to be used for a subsequent method invocation corresponding to the `ServiceRequest`. |
| 13, 14. `ServiceRequest` | CORBA method. By invoking this method on the `SSFAccessManager` object, which in turn relays the same invocation to the `IMRService` object, the client (`AccessPointToSL`) transmits necessary service parameters to the `IMRServiceLogic`. |
| 15. `TranslationRequest,` 16. `[E164SEN]` | SEN internal communication between SCF and SRF functionality. On reception, the `TerminalProfile` object returns the Broker E164 number `[E164SEN]` to the `IMRService` object. |
| 17, 18. `RequestReportSSMChange` | CORBA method. Through the mediation of the `SSFAccessManager` object, this method is invoked by the `IMRService` object (client) on the `AccessPointToSL` acting as a server, in order to instruct the B-SS&CP to monitor for SSM state changes, and consequently for service session related events. |
| 19. `PutInMonitoringList,` 20. `SetCallEvent,` 21 `CCF_ArmDP_Invoke` | B-SS&CP internal communication messages between SSF and CCF functionality, instructing the last one that session related events (which are mapped into call related events from the CCF) has to be reported accordingly to the IMR Service Logic. |
| 22,23. `JoinPartyToSessionAnd-LinkLegToBearer` | CORBA method. Through `SSFAccessManager` object, this method is invoked by the `IMRService` object (client) on the `AccessPointToSL` object, in order to request the SSF functionality to join a new party (that related to the Broker) to the service session. This means instructing the B-SS&CP to establish the bearer connection with the Broker. |
| 24. `CreateNewPartyAndLeg,` 25. `RouteConnection,` 26. `CCF_CompleteBearer_Invoke` | B-SS&CP internal messages exchanged between SSF and CCF functionality, communicating (24) to the first one to create a new Party (the Broker with associated E164 number) to be managed by the Inner SSM object and instructing the last one to route (25) the connection towards the Broker via Signaling (26). |
| 27. `SETUP` | Signaling message sent to the SEN/Broker (Called Party) by the B-SS&CP Signaling modules (instructed by the `CallControl` object). |

| Message | Action |
|---------|--------|
| 28. `Connect` | Signaling message sent by the Called Party to the B-SS&CP Signaling modules in order to indicate call acceptance. For the sake of simplicity, the exchange of `Call Proceeding` message preceding the Connect (28) is not shown. |
| 29. `CCF_DP_Signal,`<br><br>31. `SSM_State_Change` | B-SS&CP internal communication message between CCF and SSF functionality indicating to the last one that an event (connection established) detected by the `CallControl` object has to be handled by the Inner SSM object through an `SSM State Change` (31). |
| 30. `Connect` | Signaling message sent by the B-SS&CP Signaling modules to the Calling Party/STB in order to indicate call acceptance. For the sake of simplicity, the exchange of `Call Proceeding` message preceding the Connect (30) is not shown. |
| 32. `ReportBearerConnStatus` | Internal communication message between the Inner SSM object and the `AccessPointToSL` instructing the second one for reporting to the IMR service object that the event STB/SEN bearer connection established was verified. |
| 33, 34. `Report_SSM_Change` | CORBA method. Invoked by the `AccessPointToSL` object on the IMR service (through the mediation of the `SSFAccessManager` object), this method reports to the SEN for service session related events. In this specific case, the STB/SEN connection establishment is reported. |

## 10.3   BVT Service with Mobility Management Support

The UML methodology has also been adopted to describe the specific behavior of the *Broadband Video Telephony* (BVT) service with *Mobility Management* (MM) support.

Broadband Video Telephony is an interactive service that belongs to the class of conversational services. It is a real-time, multimedia, two-party service that provides two geographically separated users with the capability of exchanging high quality voice information, together with the transmission of high quality video.

Within the MARINE framework, the BVT service is provided within an environment, where broadband fixed terminals have been made available and personal mobility is supplied.

In fact, for example, it is possible for the user to acknowledge his position to the network through *registration* procedures provided by the Mobility Management functionality.

## 10.3.1 Mobility management

Mobility Management relates to modifications of service and subscriber related data that are usually handled off-line (i.e. not synchronously with an actual invocation of the service). A huge array of such data has to be maintained and updated properly in response to subscriber roaming, terminal registration, paging and changing of user preferences. Such actions can be viewed as operations on a database. Reflecting the practicalities of large telecommunication systems, a distributed approach is always taken in designing this database by focusing on specific entities as *Home Mobility Provider* (HMP), *Roaming Broker* (RB), *Serving Network Mobility Provider* (SNMP), and so on. These entities can abstractly be viewed as nodes residing at different levels of the conceptual hierarchy of the distributed database. While the scheme is generic enough to facilitate arbitrary degrees of information distribution, a reasoned choice is usually made on how this distribution should be in accordance to trade-off between complexity of the system, speed of transactions and expected levels of user and subscriber mobility. An example can be represented by a database which is usually centralized at network (national) level whilst allowing for a certain degree of information distribution in the case of users roaming on different networks.

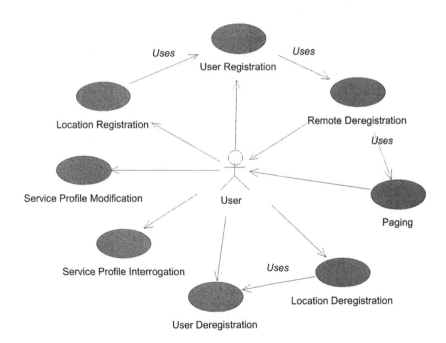

**Figure 10.3** Mobility management use case diagram

Considering the use case diagram shown in Figure 10.3, it can be seen that only one actor needs to be defined for the purposes of Mobility Management procedures, and that this is the end user. The actor can initiate the following procedures:

- Register for a service
- De-register for a service
- Register of a terminal
- De-registration of a terminal
- Service Profile interrogation
- Service Profile modification.

The use cases initiated by the system are:

- Remote de-registration
- Paging.

Before presenting the diagrams specifying a BVT service invocation, procedures for accessing, updating or removing MM information are described.

### 10.3.2  Mobility management objects characterization

For an understanding of design issues concerning MM procedures, it is necessary to introduce the main Mobility Management Objects defined for the system. The MM objects are very tightly coupled and communicate with each other through CORBA method invocations. Taking advantage of CORBA DPE functionality, MM objects residing on a specific machine can invoke arbitrary methods on other MM objects located on a different machine, in order to perform their operations in the same manner as if all the objects were residing on the same machine.

In the following, the Mobility Management objects are characterized for the MARINE framework. Some object names (`HomeMobilityProvider`, `RoamingBroker`, `ServingNetworkMobilityProvider`) and their roles have been directly derived from the *Roaming Role Model* and the *Service Provision Role Model* analyzed for UMTS (ETSI/TG.25).

**Table 10.3**   Mobility management objects

| Object | Operation |
|---|---|
| `UserAssistant` | The `UserAssistant` (UA) object contains the information related to the User's Profile, as well as registration information for services. |
| `Terminal` | The terminal objects contain information related to the actual physical terminal capabilities. The terminal object also contains a list of the current users that can be logged to a specific physical terminal. Terminal objects are created when a location registration occurs, and are destroyed when a location de-registration is completed. |
| `HomeMobilityProvider` | If a subscriber has a contractual relationship with a Mobility Provider, then by definition this is the `HomeMobilityProvider` (HMP). A roaming contract can be established between different |

| Object | Operation |
|---|---|
| | Mobility Providers, or also between the HMP and a `RoamingBroker` (see the following object), allowing (in this case) subscribers to obtain widespread mobility support through all Mobility Providers contracting with the Broker. Within the MARINE context, the HMP object maintains a list of references to the `RoamingBroker`. The HMP is always contacted when the `RoamingBroker` object ignores the location of `UserAssistant` object. |
| `RoamingBroker` | As already introduced, the RB may establish a roaming contract with different Mobility Providers. The term 'broker' implies signaling support for mobility and charging information. In MARINE, the `RoamingBroker` maintains a list to the SNMPs: `ServingNetworkMobilityProvider` (see the following object description). |
| `ServingNetwork-MobilityProvider` | In order to supply Mobility to its subscribers, a Mobility Provider may have a contractual relationship with one or more Serving Networks. In the MARINE framework, the SNMP object manages a list of User Assistants that are registered for the selected services (e.g. BVT). The `ServingNetworkMobilityProvider` routes the service requests to the involved objects in order to accomplish the Mobility Management procedures. For occasions where the SNMP does not know the reference of a `UserAssistant`, it forwards these requests to the `RoamingBroker`. In case the `RoamingBroker` also does not contain the above-mentioned reference, it sends such requests to the HMP (with whom the user has a contractual relationship). |
| `MobilityKeyHandler` | The `MobilityKeyHandler` selects the correct SNMP to send a request onto for the user. |
| `M-CUSF` | The object named Mobile Call Unrelated Service Function handles (`M-CUSF`) all the B-SS&CP call unrelated events that are associated with the Mobility Management procedures. The `M-CUSF` object is thus responsible for interpreting messages coming from the Call Unrelated Signaling and invoking methods on the `MobilityKeyHandler`. For the sake of simplicity, to minimize the number of interactions in related collaboration diagrams, the M-CUSF encompasses two objects: `AccessPointToMobilityKeyHandler` and `AccessPointToMobilityServiceLogic` which play similar roles to those described for 'peer' objects (`AccessPointToSKH`, `AccessPointToSL`) explicitly defined for the IMR and BVT service. |

| Object | Operation |
|--------|-----------|
| MBCUSMAccessManager | The MBCUSMAccessManager can be considered as an object playing an intermediary role between the M-CUSF and M-SCF functional entities. Each method invocation exchanged between the M-CUSF and the MKH and between the M-CUSF and the SNMP is processed by the MBCUSMAccessManager. |

As far as previously described for the SSF Access Manager object, the MBCUSM Access Manager also acts as an intermediary by relaying CORBA method invocations between different Functional Entities. Thus, the interaction between the M-CUSF object and MBCUSMAccessManager object (and in a similar way the interaction between SSFAM, AccessPointToSKH, AccessPointToSL) acts as a sort of 'proxy' necessary to mediate between the message-based protocols that operate along the Terminal-B-SS&CP path and method invocations used between objects implementing CORBA functionality along the B-SS&CP-SEN path.

### 10.3.3  Alternative MM-information distribution schemes

As introduced in the previous sections, a number of alternative ways to distribute the MM information among the nodes of a conceptual hierarchy comprising UA, RB, HMP and SNMP exist. Of particular significance in determining the degree of distribution are the *Options* available for the UserAssistant object (which also has an important impact on the Mobility Management collaboration diagrams). In fact, depending on the User Assistant option adopted, different types of information should be stored at each node.
The following distribution schemes can be considered:

1. Single UserAssistant for all registrations, which means: one, single, global User Assistant contains the registration information for all service registrations.
2. UserAssistant per SNMP: one User Assistant object is instantiated on every SNMP at which the user is registered for a service.
3. UserAssistant per service: for every service registration a UserAssistant object is instantiated. Thus, at any one time there could be multiple User Assistants (even on the same SNMP) representing the user.

Option 2: UserAssistant per SNMP, represents a reasonable trade-off between the conflicting goals of reducing node (processing) and network (signaling) load. The rational behind the adoption of this option is as follows:

- Adopting Option 1 would result in a very 'centralized' approach to mobility management. It must be remembered that one of the main requirements of the Mobility Management procedures is the de-centralization/distribution of the registration information for each user.
- Adopting Option 3 would result in a large number of user assistants on each SNMP. This could lead to performance and load problems on each node.

Option 2 allows for the distributed decentralization of the Mobility Management information, while minimizing the required number of user assistant objects necessary for management of the information.

Consequently, for the specification of the Mobility Management and BVT procedures that follow, it is assumed that this option is chosen.

### 10.3.4 MM collaboration diagrams

Instead of showing in detail the collaboration diagrams for all MM related procedures, we describe only the relevant scenario for the User Registration MM procedure. This particular procedure is executed whenever a user is registered for a specific service (BVT) in the network, whose domain is represented by the SNMP.

Depending on the type of BVT service calls the user wishes to make, three registration possibility are given:

- *In-Call Registration*: incoming calls only
- *Out-Call Registration*: outgoing calls only
- *In and Out-Call Registration*: both incoming and outgoing calls.

The In-Call Registration feature enables the user to register from the current terminal address for incoming calls to be presented to that terminal address.

The Out-Call Registration feature enables the user to register for outgoing calls from the current terminal address. The In- and Out-Call Registration obviously encompasses the previously mentioned features.

It is important to highlight that the User Registration procedure is mandatory, because a user needs to have a *Temporary Mobile User Identity* (TMUI) for setting up a BVT service call.

In more detail, the following definitions for users and terminal numbering would help in clarifying this topic:

- *TMUI: Temporary Mobile User Identifier* – this identifier is used by the network for the association between a mobile user and the terminal he is currently using. The TMUI is an identifier that is assigned to the user.
- *TMTI: Temporary Mobile Terminal Identifier* – this identifier is also used by the Network for the association between a mobile user and the terminal he is currently using. The difference is that the TMTI is assigned to the terminal.
- *PoA: Point of Attachment* – this number is the physical address of the terminal according to the numbering plan of Public Networks (E164 number).

Thus, to have the possibility to perform a BVT service call, the above-mentioned identifiers must be known by the network. In particular, a successful User Registration procedure will return the TMUI/TMTI values assigned to the user/terminal, while the Point of Attachment will be returned by a successful result on invoking the Paging Procedure (see the following sections).

User Registration can provide more details about the way in which UA, SNMP, RB and HMP co-operate in providing a distributed database service to the user and to the network.

The collaboration diagrams in Figure 10.4 covers the scenario for user registrations where the user does not have any registrations for a specific service anywhere. After a User Assistant has received a User Registration message and it has performed some information clean up, it will migrate to the requesting SNMP. This operation will involve the cloning of the current User Assistant to contain the user profile for the requested service, followed by the migration of that clone to the requesting SNMP.

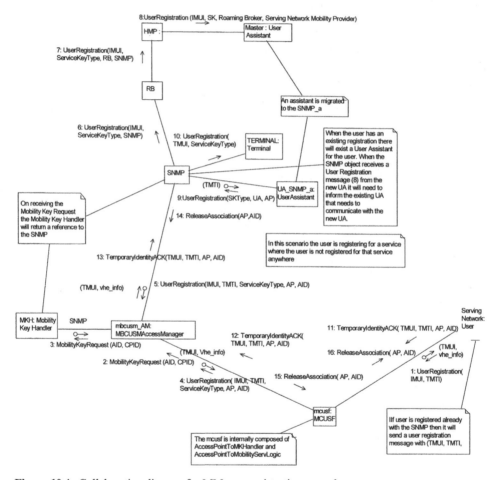

**Figure 10.4** Collaboration diagram for MM user registration procedure

**Table 10.4** Actions performed by the objects involved during user registration procedure (for the unsuccessful case flows 14-16 are relevant)

| Message | Action |
| --- | --- |
| 1: | The user (serving network user) initiates a registration procedure by sending a user registration request (Call Unrelated signaling message) towards the network. This message is processed by the MCUSF object residing on the B-SS&CP. |

| Message | Action |
| --- | --- |
| 2: | The MCUSF object issues a `MobilityKeyRequest` (CORBA method invocation) towards the `MBCUSMAccessManager` objects. |
| 3: [SNMP] | The `MBCUSMAccessManager` relays this method invocation towards the `MobilityKeyHandler` object which returns a reference [SNMP] to the SNMP object to be used for this registration request (SNMP a). |
| 4: | The MCUSF object invokes a User registration request (CORBA method invocation) on the `MBCUSMAccessManager`. |
| 5: | The `MBCUSMAccessManager` relays this method invocation to SNMP a. |
| 6: | Considering the initial hypothesis which foresees the user not registered at the SNMP for any service, this object will send the user registration request to the `RoamingBroker` associated with the SNMP a. |
| 7: | The `RoamingBroker` object is unaware of any SNMP where the user is registered for the service. It will then forward on the user registration request to the `HomeMobilityProvider`. |
| 8: | The `HomeMobilityProvider` will send the message to the `MasterUserAssistant` that manages a list of `UserAssistant` objects associated to specific services. |
| 9: | Since the user is not registered for the BVT service, the `MasterUserAssistant` will instantiate a `UserAssistant` object for that service for communicating registration information to the requesting `ServingNetworkMobilityProvider`. As already mentioned, this operation will involve the cloning of the current `UserAssistant` to contain the user profile for the requested service, followed by its migration towards the location of the requesting SNMP (SNMP a). Once migrated, the `UserAssistant` Agent will issue a user registration message back to SNMP a. |
| 10: [TMTI] (TMUI, vhe_info) | The SNMP will in turn issue a user registration message to the terminal object, in order to store registration information about the user and the service. As a result of this operation the SNMP returns the TMTI value to the User Assistant for message 9. In addition, the SNMP returns back the TMUI value and VHE (Virtual Home Environment) information to the (serving network) user. |
| 11: | To complete with success the User Registration procedure user sends a `TemporaryIdentityACK` (Call Unrelated Signaling message) to the MCUSF, in order to ask the Network/SNMP to acknowledge the registration with the associated TMUI/TMTI values. |
| 12: | The MCUSF object issues a method invocation (corresponding to the `TemporarIdentityAck` message) on the `MBCUSMAccessManager`. |
| 13: | The `MBCUSMAccessManager` relays this method invocation to the SNMP. |
| 14 | If an error occurred or the SNMP had not received a `TemporarIdentityAck` message within a certain time frame, then a ReleaseAssociation message is sent to the `MBCUSMAccessManager`. |
| 15: | The `MBCUSMAccessManager` will forward the `ReleaseAssociation` message to the MCUSF object. |

| Message | Action |
|---------|--------|
| 16:     | The `MCUSF` will send the `ReleaseAssociation` message to the (serving network) user |

### 10.3.5  BVT service description

Having described the User Registration procedure and gained an understanding of the roles played by MM objects such as the RB, HMP and so on, it is now possible to examine a typical invocation of the BVT service with Mobility Management support.

Figure 10.5 reports the use case diagram for a BVT service invocation. It can be seen that the BVT call setup procedure is initiated by the calling user, and may use the *Paging* procedure to locate the *Point of Attachment* of the called party.

A large number of different scenarios can be applicable for each BVT service session, depending on factors such as the network to which the calling and the called user are registered and the mechanisms employed for object interactions.

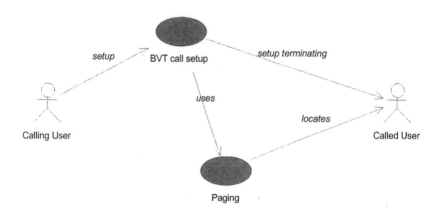

**Figure 10.5**  Use case diagram for a BVT service invocation

With reference to the first factor, the scenario where the calling and called users are registered in different networks (inter-switch scenario: two different B-SS&CP) will be presented. The corresponding collaboration diagrams will be shown. Due to the complexity of the scenarios, the collaboration diagrams will be decomposed into two parts:

- *Connection Setup*. This collaboration diagram details the object interactions necessary to set up an end-to-end connection between two users. The *BVT Service* object invokes the *Get Routing Info* method on the *BVT Assistant* object. The *BVT Assistant* object returns the routing information to the *BVT Service* object.
- *Determine Routing Information*. This part describes the determination of the routing information for the called user (Paging Procedure).

The Connection Setup part foresees a successful result on determining the Routing information.

The following objects have been characterized as specific for the provision of the Broadband Video Telephony service:

**Table 10.5** Broadband video telephony service objects

| Object | Operation |
|---|---|
| BVT Service | The BVT Service object represents the open BVT Service between end users. In addition, it contains the logic necessary to communicate with the Route Finder in order to determine location information for the remote party. |
| Route Finder | The Route Finder is responsible for determining the called party location for the BVT Service in a BVT call. The Route Finder performs the functionality of the BVT User Assistant. |

### 10.3.5.1 BVT connection setup

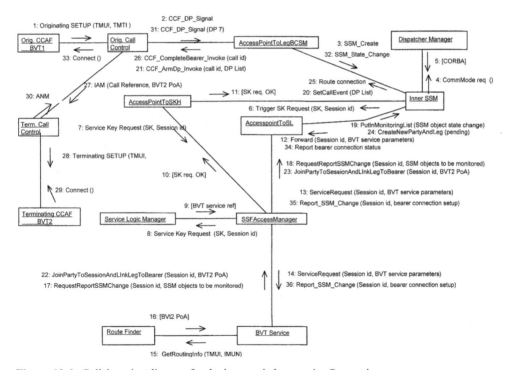

**Figure 10.6** Collaboration diagram for the inter-switch scenario. Connection setup part

Figure 10.6 reports the collaboration diagram for the connection setup part in case the calling and called users are registered on different networks.

As already introduced, for realizing a service session the BVT (like IMR) requires a Call Establishment between a Calling (BVT Terminal 1) and a Called party (BVT

Terminal 2). In fact, most of the BVT Connection Setup scenario shown into the diagram, relies on similar objects and interactions already described in Section10.2.2.

**Table 10.6** Actions performed by the objects when BVT connection setup is involved

| Message | Action |
|---|---|
| 1. OriginatingSETUP | UNI Signaling message sent to the first B-SS&CP by the Call Control functionality (Orig. CCAF) residing on the BVT 1 Terminal (Calling Party). |
| 2. CCF_DP_Signal,<br><br>3. SSM_Create | B-SS&CP internal communication mechanism between CCF and SSF functionality indicating (2) to the last one that an IN Trigger for a BVT service request detected by the CallControl object has to be handled by the InnerSSMobject. For this purpose, this object creates (3) and manages an Switching State Model (SSM) data structure to handle (through opportune SSM state changes) all service session related events. |
| 4. CommModeReq,<br><br>5. [CORBA] | B-SS&CP internal communication between the SSF functionality and the DispatcherManager object in order to identify through an opportune request (4) the communication mode to be established with the SEN equipment; via CORBA (5) or via the INAP protocol. Within this context, CORBA communication is considered. |
| 6. TriggerSKRequest | B-SS&CP internal communication between the InnerSSM object and B-SS&CP objects implementing CORBA functionality (AccessPointToSKH, AccessPointToSL). This message activates the AccessPointToSKH object for a subsequent method invocation corresponding to the ServiceKeyRequest. |
| 7, 8. ServiceKeyRequest | CORBA method invocation. By invoking this method on the SSFAccessManager server object the client (AccessPointToSKH) initiates a BVT service session including the appropriate ServiceKeyNumber and session identifier (SessionId) as method parameters. The SSMAccessManager object in turn, acting as a client relays this method invocation towards the ServiceLogicManager object. When this method returns, the SSFAccessManager maps ServiceLogic instance to SessionId. The ServiceLogicManager generates a new BVT Service instance by cloning a mobile agent that serves as a prototype implementation for that service. |

| Message | Action |
|---|---|
| 9. [BVTServiceRef],<br>10, 11. [SKReqOK],<br>12. Forward | B-SS&CP internal communication. After created the Service instance the ServiceLogicManager returns BVTService object reference to the SSFAccessManager. On successful invocations [SKReqOK] the InnerSSM object Forwards SessionId and necessary BVT service parameters to the AccessPointToSL object to be used for a subsequent method invocation corresponding to the ServiceRequest. |
| 13, 14. ServiceRequest | CORBA method. By invoking this method on the SSFAccessManager object, which in turn relays the same invocation to the IMRService object, the client (AccessPointToSL) transmits necessary service parameters to the BVTServiceLogic. |
| 15. GetRoutingInfo<br>16. [BVT2PoA] | This invocation activates the RouteFinder for locating the Called UserAssistant in order to identify the Routing information [BVT2PoA] associated to the Called Party. |
| 17, 18. RequestReportSSMChange | CORBA method. Through the mediation of the SSFAccessManager object, this method is invoked by the BVTService object (client) on the AccessPointToSL acting as a server, in order to instruct the B-SS&CP to monitor for SSM state changes and consequently for service session related events. |
| 19. PutInMonitoringList,<br>20. SetCallEvent,<br>21 CCF_ArmDP_Invoke | B-SS&CP internal communication messages between SSF and CCF functionality, instructing the last one that session related events (which are mapped into call related events from the CCF) has to be reported accordingly to the BVTService Logic. |
| 22, 23. JoinPartyToSession-AndLinkLegToBearer | CORBA method. Through SSFAccessManager object, this method is invoked by the BVT service object (client) on the AccessPointToSL object, in order to request the SSF functionality to join a new party (the one related to the BVT 2 terminal) to the service session. This means instructing the B-SS&CP to establish the bearer connection with BVT2. |
| 24. CreateNewPartyAndLeg,<br>25. RouteConnection,<br>26. CCF_CompleteBearer_Invoke | B-SS&CP internal messages exchanged between SSF and CCF functionality, communicating (24) to the SSF to create a new Party (the BVT2 with associated PoA) to be managed by the InnerSSM object and instructing the CCF to route (25) the connection towards BVT2 via Signaling (26). |

| Message | Action |
| --- | --- |
| 27. IAM | Initial Address Acknowledge. NNI Signaling message (handled by the `TerminatingCallControl` object located on the second B-SS&CP) sent to the second B-SS&CP by the first B-SS&CP for setting up the BVT Call. |
| 28. TerminatingSETUP | UNI Signaling message sent to the Call Control Functionality residing on the BVT2 Terminal (`TerminatingCCAF BVT2`) by the second B-SS&CP (through `TerminatingCallControl` object). |
| 29. Connect | UNI Signaling message sent by the BVT2 Terminal (through its CCAF functionality) towards the second B-SS&CP in order to indicate BVT call acceptance. For the sake of simplicity, the exchange of `Call Proceeding` message preceding the Connect (29) is not shown. |
| 30. ANM | Answer Message. NNI Signaling message sent to first B-SS&CP by the second B-SS&CP in order to indicate BVT Call acceptance. |
| 31. CCF_DP_Signal<br>32. SSM_State_Change | B-SS&CP internal communication message between CCF and SSF functionality indicating to the last one that an event (`Connection Established`) detected by the (Originating) Call Control object has to be handled by the `InnerSSM` object through an `SSM_State_Change` (32). |
| 33. Connect | UNI Signaling message sent by the B-SS&CP 1 Signaling modules to the Calling Party/BVT1 in order to indicate call acceptance. For the sake of simplicity, the exchange of `Call Proceeding` message preceding the Connect (33) is not shown. |
| 34. ReportBearerConnStatus | Internal communication message between the `InnerSSM` object and the `AccessPointToSL` instructing the second one for reporting to the BVT service object that the event: 'BVT1/BVT2 bearer connection established' was verified. |
| 35, 36. Report_SSM_Change | CORBA method. Invoked by the `AccessPointToSL` object on the BVT service (through the mediation of the `SSFAccessManager` object), this method reports to the SEN for service session related events. In this specific case the BVT1/BVT2 connection establishment is reported. |

### 10.3.5.2  Determine routing information

Figure 10.7 shows the collaboration diagram for the BVT Assistant locating the called party when the called party is registered in a different network (and RB) with respect to the one of the calling party. The BVT Assistant interrogates the SNMPo and RBo, (o – originating) RBo interrogates the HMPt, (t – terminating) the HMPt interrogates the RBt, and the RBt interrogates the SNMPt.

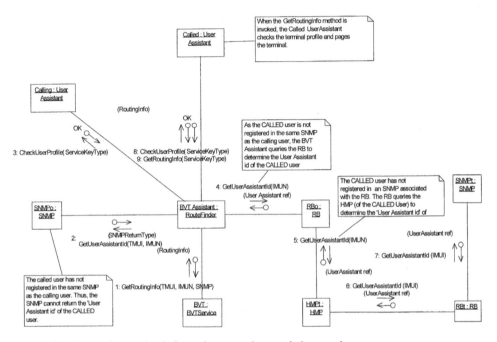

**Figure 10.7**  Determine routing information part – inter-switch scenario

**Table 10.7**  Actions performed by the objects when determining routing information – Inter-switch scenario

| Message | Action |
|---|---|
| 1. GetRoutingInfo | The GetRoutingInfo is invoked by the BVTService object on the RouteFinder in order to determine the routing information/PoA of the called user. |
| 2. GetUserAssistantId | On invocation, SNMP originating returns: |
| | UserAssistant for calling user, |
| | RBo: the called user is not registered in this SNMP – thus, RouteFinder is given the reference of Roaming-Broker associated with this SNMP, allowing RouteFinder to query the RoamingBroker for called UserAssistant. |

| Message | Action |
|---------|--------|
| 3. `CheckUserProfile` | On invocation, calling `UserAssistant` checks if calling user is allowed to invoke the service. |
| 4. `GetUserAssistantId` | On invocation, `RBo` invokes a similar method on `HMP` terminating. |
| 5. `GetUserAssistantId` | On invocation, `HMPt` invokes similar method on `RBt`. |
| 6. `GetUserAssistantId` | On invocation, `RBt` invokes a similar method on `SNMPt`. |
| 7. `GetUserAssistantId` | On invocation, `SNMPt` returns a reference to the Called `UserAssistant`. |
| 8. `CheckUserProfile` | On invocation, Called `UserAssistant` checks if the called user is allowed to use the service. |
| 9. `GetRoutingInfo` | On invocation, the Called `UserAssistant` returns (through the `RouteFinder`) the Routing information (`RoutingInfo`) to the `BVTService` object. |

# References

Booch, G. (1994) *Object-Oriented Analysis and Design with Applications*, Second Edition. Addison-Wesley.

Coad, P. and Yourdon, E. (1991a) *Object-Oriented Analysis, Second Edition.* Yourdon Press Computing Series.

Coad, P. and Yourdon, E. (1991b) *Object-Oriented Design.* Yourdon Press Computing Series.

ETSI, Service Mobile Group SMG1-UMTS, Memorandum of Understanding GSM MoU TG.25.

Jacobson, I., Christerson, M., Jonsson, P. and Overgaard, G. (1992) *Object-Oriented Software Engineering: A Use Case Driven Approach.* Addison-Wesley.

OMG, Specification ad/97-08-02 (1997) UML version 1.1.

OMG, Document ad/99-06-08 (1999) UML version 1.3.

Rumbaugh, J. and Blama, M. (1991) *Object-Oriented Modeling and Design.* Prentice-Hall.

# Index

Printed and bound by CPI Group (UK) Ltd, Croydon, CR0 4YY

27/10/2024

14580295-0001